MATT J PIKE

Edited by Scott Moore / K.E. Fraser

Cover illustrations: Steve Grice
Cover design: Matt J Pike

****

****

This novel is entirely a work of fiction. The names, characters and incidents portrayed in it are the product of the author's imagination. Any resemblance to actual persons, living or dead, or events or localities is entirely coincidental.

Published by Zombie RiZing Books

Paperback Edition: November 5, 2017
ISBN: 978-1-64204-947-3

www.mattpike.co

Cover design: Matt Pike/Steve Grice

BOOKS BY MATT J PIKE

Apocalypse Survivors:

JACK
(Prepping/adventure)
Apocalypse: Diary of a Survivor
Apocalypse: Diary of a Survivor 2
Apocalypse: Diary of a Survivor 3
Apocalypse: Diary of a Survivor 4

NORWOOD
(Dark action/thriller)
The Parade
War Parade
Reign on the Parade
Hit Parade COMING 2026
The Last Parade COMING 2027

*

Next Level: The Robot AI Zambie End Game
(Zombies meets robots sci-fi horror comedy)

Zambies!
Getting Plutoed
Life on Twearth

Hart & Sol
(Totally inappropriate sci-fi comedy)
(co-written with Russell Emmerson)

Starship Dorsano Chronicles:
(YA Inappropriate sci-fi comedy)

Kings of the World
War & Quel
Revenge & Qualia COMING 2027/28

*

Zombie RiZing:
(Middle grade epic fantasy horror comedy)

The Beginning
Dreeks' Horde
Dragon's Wrath

Death's Door
Dark Abyss
Castle Death COMING 2027

*

****

I would like to thank a number of people, all of whom have helped in parts, big and small, to turn this book into a reality:

Scott Moore, Lisa Chant, Russell Emmerson, Steve Grice, Katie Lowe, Wayne Bosch, Tory Bosch, Jan Pike, Anna Pike, Lisa Smth, Noami Jellicoe.

I would also like to thank my children – Sophie, Sam and Abby – for their constant inspiration.

****

PREVIOUSLY ON
STARSHIP DORSANO
CHRONICLES

(To be read in a booming TV announcer's voice; picture the 'previously on AMC's the Walking Dead' guy. His vocal mix of horror-action meets cigarettes, moonshine and the distress of uncomfortable footwear totally kicks ass and this type of tone and timbre would pepper the ensuing words with the cherry-ripe mix of gravitas and foreboding. Or you could just read – like, normally and stuff. I get it, you don't have to play along, leave the words up to the author and the interpretation up to the reader – that's fine. Fair call. I'll just be over here, back in my author's box, doing words, if you need me for anything.) …

*

It was just another day for Cooper Simpson and his reality-challenged, game obsessed best mate The Ginge, until a fight with local thug, the goatee-growing goliath Knuckles, and his pasty, weirdo friend Pete. A freak victory, internet stardom, detention and snooping TV crews ensue, before the MMA rematch in the park turns into an all-in brawl.

What follows is the oft-told tale of teen boy threatened to be ripped apart by larger, intellectually questionable student only to have a six foot, leathery classic grey alien – Rak Divik (their soon-to-be mentor) beam down from a spaceship, stop the fight and offer them the role as representatives on the Galactic council.

And, as we all know, having experienced this well-worn story path before, the humans always accept the alien's offer.

Before they get a chance to say their goodbyes, they are on a spaceship, heading for their new home. They discover a place where, with the power of the Universal Translator (an interface with an omnipresent supercomputer) and Creator (a device that can construct almost anything one can desire) at their disposal, they have power like never before and the possibilities are endless.

It soon becomes apparent the aliens of the galaxy have been avoiding the humans from Earth for some time. Our species' reputation for violence, war, greed and various other traits most contemporary self-respecting aliens frown upon; see us considered the galactic equivalent of white trash.

It's not long before our four heroes (and I use that term very loosely... perhaps entirely inaccurately) face a short but vomit-inducing trip on the Starship Dorsano to Galactica – the political centre of the Milky Way. They find a cosmopolitan capital of politics, knowledge and entertainment where thousands of alien species gather to share ideas, air their grievances and seek to maintain the long-held peace (spoiler alert: the latter does not happen).

Cooper sets about repairing humanity's reputation, despite the best efforts of The Ginge, Knuckles and Pete, who each appear more concerned with being the first earthling to get lucky with a one of the more human-looking aliens.

Meanwhile, after numerous attempts on their lives, and several other new councillor species being annihilated, Cooper and the boys realise all is not what it seems on Galactica. Their friend Judrev Oll soon points them in the direction of the beautiful, thought-reading Terenda and they set about getting to the bottom of the killings and mystery exiled group the Fredahn. Well, to be more accurate, Cooper does – the others seem more than happy to spend their time chasing aliwomen.

Away from threat of imminent death, the boys find themselves at the social centre of the new intake councillors as they inflict Earth's games and culture on the vulnerable intergalactic newbs. Without being aware of the impact they are having, the humans soon find themselves with heavy voting influence over a key block of councillors (the new intakes) at a crucial moment in Galactica's long history.

Their influence penetrates beyond the political – socially the boys become A-listers. Fame they let go straight to their head, culminating in a particularly awkward night where they each manage to take galactic relations to Bill Clintonesque levels. Needless to say, results vary, from high-five to hide-face.

But as morning dawns and cold reality sets in, they discover it's not only their pulling reputation that is hanging by a thread. The longstanding peace in the galaxy threatens to explode as political dynamite collides – it's the establishment (Big Three) versus the new power (Dihflua faction), with the humans, and the voting block they didn't even realise they were leading, left right in the middle, holding the match in a room full of fuses.

The boys' mentor and Big Three bigwig, Rak Divik, proves to be anything but the peaceful, gentle grey he first appeared and soon is enemy no. 1. It all reaches a head the night Knuckles finds out he is going to be a father (see hide-face, two paragraphs above) and Pete oversees a musician massacre (not his finest managerial moment). Rak Divik makes his move on the humans –The Ginge is gunned down and Cooper is kidnapped, his broken body dumped on a cold and hostile moon.

All seems lost, doubly so when Knuckles and Pete realise it's up to them to save the day. With the help of Judrev, they save Cooper from geological terror, life-threatening injuries and a fleet of spaceships, then discover Terenda has brought The Ginge back from the dead. The gang reunite to plan the most politically explosive vote the council has ever seen.

Their enemy is on to them and Rak Divik, and his Big Three forces send an army to stop the boys reaching the council chambers. All seems lost when the boys' new intake comrades join their side and, armed with The Ginge's mega-weapon – The

Ginge-O-Tine, fight their way to the chambers and their right to be heard.

After a brutal battle, the surviving members of the new intake army reach the chambers. The councillors of the galaxy watch a movie Pete has created about the boys' time on Galactica, incorporating proof of the underhanded work of the Big Three, before Cooper presents to the council. His proposal (to abolish the longevity vote) passes and the power of Rak Divik and the Big Three is crushed forever. A defeated Rak Divik goes fifty shades of cray-cray and tries to kill Cooper, only to discover the boys' power and knowledge of the UT is no longer newb status and he is, literally, fried at the hands of The Ginge.

In a last act of defiance, Rak Divik sends a fleet to destroy the Earth but is thwarted at the last minute by the boys and their now long list of galactic allies. Cooper, The Ginge, Knuckles and Pete head home as heroes of the galaxy and soon learn their abduction (acquisition) to Galactica had been caught on tape – thanks to a local TV crew, the world knows their story. They return famous beyond words. Their tale, and more importantly, title as world galactic leaders, had not gone unnoticed in political circles, with Earth's leaders none too pleased with the boys' promotion from useless teenagers to planet leaders.

The triumphant teens set course for the Sydney Opera House, a date with fame and a showdown with the Earth's political heavyweights.

****

PART 1:
HOME

CHAPTER 1

"They outnumber us, maybe 15 or 20 to one," screamed The Ginge as he scanned the skies from the turret viewfinder. "We're screwed. We should've just laid low behind that moon until they got closer. Whose stupid idea was it to jump out in attack mode?"

Everyone looked at Knuckles who was sitting at the other turret. "What? It looked clear enough to me."

"Clear enough?" said The Ginge. "That's clear enough? I can barely see the mothership through all the fighters. And what's that little dot of black way back there in the distance? Oh, that's space. I can see a gazillion attackers, bits of mothership and a speck of space. Brilliant."

"Shut up, Ginge. They came out of nowhere."

"Shut up the lotta ya," said Pete as he struggled at the controller to steer their ship around the first barrage of enemy fire.

"Pete's right," said Cooper, "we've got to think of something and fast. Anyone?"

"We should aim for the small speck of black and warp our asses out of here."

"Run away?" said Knuckles. "You want to run away?"

The ship shuddered as it took a direct hit from enemy fire, the boys struggled to keep their stations.

"Sounds good to me," said Pete.

Knuckles fired a stream of pulse fire from his turret into a bogey ship as it fanned past their craft. "Pussies."

"No one's running anywhere," said Cooper, "we can still stop them. I think we need to hit them at the mothership, that way we can take the attack ships out of play…"

"Whoa, whoa, whoa," said Pete, "tell me you're not serious."

"Deadly serious. We need to draw them out further. Pete, bank this thing around and get 'em chasing us."

Pete complied. "This is the stupidest plan ever. Why don't I get to make the decisions?"

Pulse fire whizzed past the ship at all angles; the hull took a couple of direct hits. "I mean I'm the pilot, I'm the one who—"

"No one cares, Pete," said The Ginge as he took out an enemy trailing off the starboard side. "Boom!"

"Nice one, Ginge." Cooper looked at the chasing attackers. "You lads think you can hold 'em off for a minute more?"

Knuckles shot a wave of pulse fire, clipping two other attackers. "Do we have a choice?"

One of the injured attack ships ploughed into another and incinerated the pair.

"No," said Cooper. "Pete, how are your slow switching skills these days?"

"Pretty good… ahh, I see what you're doing."

"On my mark, hit a slow switch, flip the ship around, and drop us just short of the mother ship."

The ship shuddered again.

"Nice! On it."

"Eight o'clock – another squadron," said The Ginge as he ploughed a line of pulse fire at the new threat.

"I've got my own problems over here." said Knuckles as enemy craft gathered in greater numbers.

The ship took another hit, which shuddered the hull and sent everything not strapped down on a dance around the cockpit.

"I'm ready," said Pete, "now?"

"Just a little bit more," said Cooper.

Pete looked at Cooper as his hands hovered over the controls. "Now?"

"Just a little bit more."

"Now?"

"Just a li—"

"Will you two shut up," said The Ginge. He let out an extra intense round of pulse fire to vent. "It's like a broken record."

"What's a record?" said Knuckles.

Another wave of fire got through the ship's defences and hit the hull. Cooper studied the damage readings – they were safe, for now. "Hang in there a little bit more," he said.

Pete paused for what he thought seemed like ample time to make a new enquiry, "Now?"

He was alone in that belief.

"Really? You're gonna start doing that again?" said The Ginge. "Could you be any more irritating?"

"Shut up, Ginge" said Cooper. "Nearly there Pete."

The pressure, the craziness, the moment – time fired at high speed yet seemed to stand still. Cooper studied the battlefield from his heads-up display (HUD), Pete weaved the craft through the safest spaces he could find, Knuckles and The Ginge repelled enemy fire and craft. Everyone knew what was coming next and nothing but the passing of time could make it happen. And like time itself, nothing could stop it…

"Now?" said Pete.

Knuckles let out a frustrated roar.

"Seriously?" said The Ginge. "You seriously said that again?"

Knuckles' roar finally drained out. He turned around to face Pete. "You are a total, full blown, di—"

The ship shuddered again, the boys knew it was a big hit. This was getting seriously dangerous. Cooper scanned his displays. "Now!"

Pete triggered the combo to set the slow switch in motion. The boys watched as the ship transformed around them. They braced for the drunken, starry sensation of being transported through space, in a nanosecond, to another location. When the main remnants of the daze echoed away, they stumbled to get their bearings. The outer shell of the ship reformed around them. They were soon back as one – ship and crew – facing the mothership. The chasing attackers miles in their wake and heading in the wrong direction.

"Brilliant work, Pete," said Cooper.

"It was your good call on the switch," said Pete.

The Ginge made a dry-retching noise. "Get a room, you two."

"Righto then," said Cooper, "Aim everything you've got at the bridge."

Knuckles and The Ginge responded with a volley of pulse fire that made the ship rattle. Within seconds the attack was met with enemy fire on a scale larger than they could muster… or handle.

Pete steered the ship as best he could past the heaviest incoming pulses. "We're taking a pounding, and the attackers are heading back this way. We can't have more than a minute before we're in their range again."

"There are more bogies heading out from the mothership," said The Ginge as he turned his fire in that direction.

"Coops, do something!" squealed Pete.

"I'm thinking."

"Well think quicker!"

"Can we warp out of here?"

"No chance," said Pete, "way too dangerous."

Cooper studied his HUD, looking for a magic solution. None came. "The only thing I can think of is the cargo bay."

"There's no use firing down there, all their ships are in the skies."

"I meant land there."

"What?" said Pete, Knuckles and The Ginge in unison.

"If we can slide into land, we should be able to shoot down any resistance on the ground, then make our way on foot to the bridge."

"Are you mental?" said Pete. "That's a suicide run. We wouldn't last a minute inside that thing."

"That'd probably give us thirty seconds more than out here. We're sitting ducks. Unless you've got something better."

There was a long silence before Pete muttered an expletive under his breath.

"Right then, we're doing this. Pete – get us on the deck as quick as you can. Ease off the gas as we get close and we can slide the ship to a halt in the landing bay. Ginge, Knuckles – stay on the turrets – as soon as we come to a stop, unleash fury on anything that moves. As soon as you've cleared enough enemies away, I'm gonna flick the hatch so we can run for the nearest cover. We can work out the best way up to the bridge from there. Got it?"

"That's the plan?" said Pete. "Fly to the most hostile location I can think of right now then work out what to do?"

"Erm... yep."

"Seriously?" said Pete as he turned to Knuckles and The Ginge. "Surely you don't think this is a good idea."

Knuckles knew it wasn't a great idea, but that paled in comparison to someone questioning his bravery. "Quit your whinging, Pete, and get on with it."

Pete complied, after mumbling something about the lack of respect he was receiving as ship's pilot. The others rolled their eyes and prepared for the battle ahead.

"They're trying to intercept," said The Ginge, "a small squadron at three o'clock.

"Got 'em," said Knuckles.

The pair focused their pulse fire on the newest threat, taking out several attackers. One of the fallen ships tumbled out of control and towards the boys' ship.

"Incoming!" yelled Knuckles.

Pete screamed as he took the ship into a sharp upwards lift in an attempt to avoid collision. A blast of enemy pulse fire rocked the ship again, it was followed by an explosion. Pete flicked to the HUD view of the battlespace but it was too late, impact was now inevitable. Pete moved his scream up in pitch and volume.

The ships collided with a sickening noise. A piece of debris burst the front shields sending the boys flying, alarms into overdrive and the air supply seeping into the vacuum of space.

Pete recovered first, stood up again and tried to regain control of the ship. It was in a heavy spin from the impact, headed for the mothership.

The Ginge reappeared from under a pile of scattered supplies and dragged his way back to his position. "My turret's gone. I've got nothing."

Cooper and Knuckles resurfaced soon after, looking up to see the final few seconds of their ship's flight time. Their view was the wall of the mother ship with a small window of relief coming from the cargo bay door. They screamed. The Ginge screamed. Pete, still trying to wrestle their ship back on track, screamed loudest of all. This, in turn, drew the others into new levels of screamage. It's basic survival principle – if the person in charge of your life is screaming for theirs, scream louder.

The ship started levelling out as the cargo bay was coming into view. The boys instinctively leaned towards it, willing the ship on. The oxygen supply was draining, but not enough to stop the screaming.

Then came the moment of realisation: they weren't going to make it. No amount of screaming or leaning or eye-closing was going to make the ship reach the cargo bay in one, or indeed any countable number of, pieces.

Humanely, that realisation came late, it only left Pete enough time to say, "Oh, sh—. Then there was black.

*

The four boys removed their gameplay apparatus and began the exhaustive process of figuring out what went wrong. The usual protocols ensued – name-calling, swearing and threats of violence – standard post-game-fail procedure.

Cooper watched it all unfold in replay on his TV. It was the last time they'd gamed together and it had been months ago. In fact, it'd been months since they'd all been in the same place together. In a strange way he missed them – he even missed the violent threats. What was happening to him?

He walked over to the window of his house and admired the vista of endless coastline in front of him. He breathed in the remote seaside air, enjoyed his view across the Great Australian Bight and smiled to himself that he'd found a place in the world that made him content. It really was the perfect place for his hideaway – isolated and peaceful – but every now and then, like now, he missed the craziness.

Rhiannon Richardson joined him in the lounge with offerings of hot chocolate. Cooper thanked her, took a sip and returned his gaze to the soothing sea.

Rhiannon ran her eyes up and down her partner. "Missing them again, Coops?"

"What?" said Cooper, lost in his thoughts. "Oh, yeah, nah. It's nothing."

"Why don't you just go tonight? I can come with, if you want."

"It's too risky. Pete and The Ginge are already going, having three of us in the same place… that's just asking for trouble."

"Madison Square Garden, though, it will be amazing."

*

Pete Bates paced back and forward in the green room. He could hear the crowd abuzz with anticipation, which didn't do anything to ease his nerves. Neither did The Ginge, who rehearsed which one of his sexy poses he would use when seated on the couch on stage.

"Will you cut that out," said Pete, "you're driving me insane."

"You can talk, I can almost see the halo of stress glowing around you" said The Ginge, "do me a favour and relax, man."

"Relax? Have you seen how many people and cameras are out there?"

"It'll be fine. Besides, if you don't slow down you'll have paced out a marathon before you get out there."

Pete nodded, paused and took ten deep breaths. "Have you heard from Coops yet?"

"Nuh. Don't think he's coming – just you and me, dude."

"Awesome. Great. That means 50 percent more questions directed at me." He started pacing back and forth again.

"Pete, everyone's here to celebrate the movie," said The Ginge as he practiced different hand-through-hair moves in the mirror. "And us."

"You seem to be doing a good enough job of celebrating yourself these days," said Pete, as he looked The Ginge up and down then shook his head.

The Ginge admired himself in the mirror. He gazed lovingly at his designer jeans, his sunglasses and his Miley Cyrus twerktastic, t-shirt. "My moves aren't for me – they're for the fans. It's like… a gift."

"Seriously? A gift? If that's a gift I thank the lord you missed my birthday."

The Ginge missed the insult as he toyed with a few more potential poses, then ran through a variety of facial expressions – earnest, cool, suave, deep, da-man, that chick digs me and I'll consider her if she plays her cards right – all his staple expressions.

Pete couldn't look directly at him, choosing to flick through a few dot-points on the movie's production for a distraction. "Oh, did I mention the studio's hoping we can keep things as G-rated as possible tonight."

"What?"

"The studio wants us to keep things clean, for the kids."

"What do you mean 'clean'?"

"Just, ya know, avoid going into too much detail about things like, I dunno, Knuckles' dick-tricks and all the alien chicks you've been with, stuff like that."

"Are you serious? That's all anyone ever wants to know about."

"I know, just keep it G-rated is all I'm saying."

"It was a long, long way from G-rated."

"Eww. Really Ginge? That's exactly what I'm talking about. I'm close enough to vomiting through nerves as it is; don't push me over the edge!"

"What? It's true. Besides, if the fans want to know, I'm gonna tell them."

Pete sighed. "Whatever. Just, ya know, go easy."

"Anyway, since when have we been taking orders from the studio?"

"Since they've made us about a billion dollars."

"Pete, my friend, we have made them even more than that… because we are stars. Get with the times; content is king; the studio works for us. Embrace your stardom, Pete."

"Whatever."

"Besides, listen to them out there, we're the biggest thing that ever happened to them. They love us. What the worst that could happen?"

"You're right," said Pete as he practised some more breathing exercises, "What's the worst that can happen?"

***

CHAPTER 2

Cooper reflected on how much had changed since they returned from Galactica. They'd been through everything – the elation of their return and the style in which they did it – they were famous beyond words. Somehow they'd acquired a manager – Jason Blackwell, the journalist who had followed and filmed them on their fateful 'visitation' night. Jason had found his own degree of fame since reporting on the most talked about news story of all time.

Cooper wasn't even sure how it happened; everything was such a blur in those first few days. The boys and Jason just got pushed together through a national press conference, which they later found out the former reporter himself had organised.

There wasn't a moment's peace from the instant they hit the ground, there were journalists and photographers camped outside their houses. Every time they went somewhere, the media was present.

It didn't take Cooper long to hate it all.

He was alone in that thought, though – the other three boys took to their fame like seasoned pros. The Ginge was his ever cheeky self – but worse – in front of a camera, Knuckles developed a swagger of false confidence that drove Cooper insane and even Pete played up to his new rock star status – somehow his sullen look and morose attire was rebadged with a pair of sunglasses to become the new 'cool' look!

Typically, when they encountered a media scrum, most of the questions were directed at Cooper, but as the days passed he found himself drifting into the background as the other members of the UFO4, as they'd been dubbed, shone.

Jason used his media experience to organise everything; the boys became robots in a schedule. There were countless TV appearances, magazine shoots, sponsorships deals, parties, hotel rooms, groupies – it was exhausting. The quality time with family and friends he'd craved so much while on the other side of the galaxy never eventuated. It just wasn't possible in those early days.

It wasn't long before the major studios starting offering ridiculous sums of money for the rights to Pete's movie. Before Cooper knew it, the boys were on a plane to the US. That's where things really got crazy. All the big talk shows, every other form of TV imaginable, appearances at major sporting events, festivals, E3, ComicCon – day after day of new cities, hotel rooms, media and fans.

For Cooper, the pace was too much to take. He hung around long enough to do all the press with the studio that signed up the UFO4 movie, then he headed home, finally managing to spend some time with his family and friends. He moved back in with his parents – it was a short-lived experiment. He had changed. Everything had changed. After dreaming of his old life, he found he was as much a fish out of water there as he was being the famous Cooper Simpson. He needed a new normal.

Pete stayed on with the studio to rework the film for cinematic release, The Ginge stayed on the circuit, appearing on just about every TV show imaginable, as did Knuckles, for a while, before he drifted off the radar as Cooper had. Within a matter of weeks the boys had each headed in their own direction – they were in different cities, or countries, doing different things. Even the gaming sessions became a thing of the past as fame, or escaping it, engulfed them.

Then things started getting weird. The US government tried to take out an injunction against the film, claiming it represented a national security issue, while at the same time continuing its campaign to discredit the boys' claims they even went to space. They were two totally contrasting standpoints but few saw the blatant hypocrisy. Secret agencies raided the movie's production company while politicians undermined the boys' credibility at every opportunity. It was a war against them on many fronts.

It wasn't long before the Australian government followed suit, as well as any other established regime Cooper could think of. It was clear the governments were threatened, the boys' position on the Galactic Council had the potential to undermine the entire Earth-bound power base, and they knew it.

In fact, when Cooper reflected back on the course of events, politicians had kept their distance from the beginning. He should have been suspicious then, because they usually buzz like flies around local success stories. When they made their intentions clear, Cooper realised that even from the moment they arrived back on earth – it had been the world's governments versus the boys.

Fortunately for Cooper, the money started rolling in and he was able to start to build his new world on his terms. He purchased a large parcel of land in the remotest part of the world he could find – along the Nullarbor Plain in the middle of Australia's southern coast. He set about creating a secure compound to monitor the other boys and the world as well as stay in touch with the rest of the galaxy. He didn't have any direct evidence but he knew he was being watched so he tried to stay off of the grid as much as possible. With the help of the Universal Translator (UT), it was not hard. The UT gave him anything he asked for, from food, to power, to access to the internet and the ability to spy on those spying on him, but in oh so much more detail.

He tried to tell the others what was going on but they weren't having any of it – they called him paranoid, or other, less-dignified terms. They weren't letting anything derail their fame and fund gravy train.

The production of the movie went underground, spread between different production houses around the globe, acting under false movie titles to keep suspicious governments at bay and to get the studio's fingerprint off the process. Even though Cooper left most of the work to Pete, it was still exciting to be a part of some guerrilla action against the powers that be.

In hindsight, one of Cooper's best moves in the early days was instructing UFO – the ship they returned to Earth on – to go into orbit after they had landed. At the time he wasn't sure what drove him to do it beyond a feeling he didn't want UFO getting into the hands of the government, or worse. Of course, having no actual proof of UFO's existence fuelled the conspiracy theorists. Some believed the boys' return to Earth, the message, everything, was a publicity stunt for the movie. But that was a small price to pay. Once settlement had passed on his property he recalled UFO back to Earth, ensuring it was obscured from prying eyes (from the road or satellite). It was the secret weapon in his fight against Earth's governments and his direct line to the UT, information and power.

He monitored the world's political leaders through UT surveillance. He could instruct the UT to capture video or audio from any conversation, from any moment that had happened in the past. He could instruct the UT to 'follow' the chain of orders downstream to see what was executed by those who followed them out. It was only then he realised how deep the war against them was.

Again he tried to warn the others. He compiled a highlights package of the worst things the world's leaders were doing and showed it to them. The Ginge didn't seem to care, Knuckles never responded and he doubted Pete even watched it.

The UT also gave Cooper access to communications beyond Earth. He stayed in touch with many of the remaining councillors, getting constantly updated on the political climate at a galactic level. 'Uneasy' was about the most positive way he could describe it.

The galaxy, Earth governments, the boys; everything in his life seemed so unstable. It reminded Cooper of a video he'd seen on the internet once, where a room was full of loaded mousetraps, each holding a ping pong ball. Then someone dropped in one final ball – it was a chain reaction of total mayhem. That was his life was right now – a series of loaded traps just waiting for someone to throw in the ball.

Sometimes he wished he could be like the others, just burying his head in the sand and enjoy the good times but, in truth, he was happy with his lot; choosing to stay on top of everything, living in his hideaway oasis, living with Rhiannon.

Rhiannon hadn't changed a bit since school. She was perfect for him – opinionated but thoughtful, respectful but challenging. Most importantly, she made him happy, she was both a reminder of where he'd come from and where he was going. Besides, when it all got too much, one look into those inviting eyes took him to another, much better, place.

In those moments when everything got to him, though, was when it hit the hardest – it was Cooper versus the world. But even with everything he was up against he still gave himself a chance. What he lacked in political clout, manpower, resources and espionage, he could make up for between his ship, the UT, Rhiannon and the boys… at least that's what he told himself. But, no matter what he believed, how he planned or what he learned, none of it would really matter until someone threw in the ping pong ball.

He would not have to wait long.

***

CHAPTER 3

"Cooper, you wished me to notify you if there is a threat to you or the other boys," said the Universal Translator.

Cooper dropped his mug and ran to the hologram of Gandalf. "I did. What's wrong?"

"It would appear that Pete and The Ginge are in imminent danger."

"Seriously?"

"Indeed, Cooper. I have been following a series of orders from the highest stations of the US government. They mean to apprehend Pete and The Ginge at tonight's appearance at Madison Square Garden."

"No way," said Cooper as he turned to the door. "Rhiannon! It's happening."

"Way, Cooper," said Gandalf. "The CIA is preparing the details of the apprehensions at this moment. They have a team armed and ready to deploy."

"I knew they were up to something. What—"

Rhiannon's footsteps echoed down the corridor then screeched to a halt as she reached the door. "What is it?"

"Madison Square Garden," said Cooper. "We're going."

"Good call, honey. Luckily I had my red dress out on the bed just in case."

"Yeah, it's probably less a red dress situation and more an army fatigues and ammo situation."

Her smile fell away. "What?"

"Gandalf, fill us in."

The UT started reciting the events that had led it to alert Cooper. As each fact was stated, a holographic replay of that moment appeared on the floor in front of them.

"Oh my God. There's dozens of them."

"48 – all heavily armed – two helicopters, two troop trucks and a number of other vehicles. They are positioning themselves to deploy their plan at 9pm, New York time. That's 45 minutes from now."

"Oh, that's just great. Umm… can you hook up a link with Pete and Ginge? If I can get them before they get on stage we can try to sneak them out of this."

"They are already on stage." The UT displayed another holographic image next to that of the preparing CIA army, this one of Pete and The Ginge wowing the 18,000-strong crowd.

Cooper muttered something under his breath and absently messed with his hair while he tried to plot his next move. "Alright, Gandalf, can you teleport us to their change rooms backstage at the Garden? And we'll need some weapons, not the Gingotine, something lighter, handheld – pistols – two for each of us, including Ginge and Pete. And I don't want to be firing ammo around a crowd, is there some sort of pulse fire that stuns but doesn't kill?"

"That is achievable."

"Sweet." Cooper looked at the communicator on his wrist and was reminded of something else. "Can you transfer the hologram display of the CIA movements to our communicators?"

"Indeed."

"And shields, we need shields – if it does go pear-shaped, I'm not sure they'll take to kindly to us firing back – stun or not. Can we rig up the shields from the Gingotine, to, erm, I don't know, say a forearm guard or something?"

"It is done."

"Brilliant. Rhiannon, you ready?"

"I need to get changed."

"Don't have time."

"Seriously? I'm wearing tracksuit pants and slippers."

"Gandalf, can you dress us in something more contextually appropriate for kicking ass."

"Yes."

"Great. Try not to make a statement. Plenty of black. No logos."

Rhiannon shot Cooper a confused look.

"For an all-seeing, all-knowing interface to everything he seriously lacks an eye for fashion. No offence, Gandalf."

Just as the UT was telling Cooper it didn't actually get offended, both humans had an instantaneous wardrobe change – black, heavy duty, military style attire, including stun guns and forearm shields. Cooper and Rhiannon looked each other up and down, then nodded their approval.

"Not bad, Gandalf, not bad at all," said Cooper before switching his attention to Rhiannon. "You ready?"

Rhiannon took a deep breath, let the air out slowly, then nodded.

"Teleporting's a bit weird at first but it's perfectly safe."

Rhiannon nodded again.

"OK Gandalf, we're ready. You ready?"

"Calculating the teleport now. This will only take a few seconds."

"Great. Oh, before I forget, have you heard from Knuckles yet?"

"No."

"Dammit. Send him another message when we're gone, tell him what we're doing and that we need him there urgently. Oh, if anything happens to us, make sure you let Jason know and put this place on high-security until Rhiannon or I return – no one else gets in, understand?"

"Yes."

Cooper jumped up and down in anticipation of the teleport, he could see the nerves in Rhiannon's eyes. He decided distraction was the best way to deal with it.

"You know, Gandalf gave me flares once, actual bell-bottoms! That's not to mention the Celine Dion t-shirt. I was starting to think he was doing it deliberately, you know. But it looks like he's starting to learn. This is prob—"

"Ready to teleport?" said Gandalf.

"Ready," replied Cooper.

In an instant they were gone from the lounge of their southern Australian hideaway and, in that same instant, they arrived backstage at Madison Square Garden, New York.

*

Knuckles stared out of the glassless window of his hut. He gazed past his view over the village and spaceport, his eyes wandered beyond the mountains and lava and into the starry-rich night sky. The view was amazing, it beat anything the Earth could provide, a combination of no smog and Gartogia's position closer to the centre of the galaxy no doubt. But Knuckles was not in the mood for wonder. He search the skies, scanning for a little blue, Earth-sized dot that he could never hope to see.

He would kill for his communicator right now, just to know what was going on back on Earth. He cursed the stupid big night of drinking after his sons were born, then he remembered the embarrassment of waking up naked, in the village square as most of the locals were preparing for lunch. He cringed at memory of his walk of shame, naked, communicatorless, back to Wendy Wendy and the newborns.

Now he was a prisoner, well, technically not a prisoner, given he had chosen to come to Gartogia to help raise his three boys, but a prisoner of duty at the very least. He laughed, a mad laugh, at how his life had worked out, halfway across the galaxy on daddy duty with his furry four-armed partner Wendy Wendy and three rapidly growing, only somewhat human, offspring. Despite being only six Earth months old, the boys were growing worryingly strong. Knuckles guessed that by the time they were a year old, he would be powerless to stop one, let alone all three.

'Seven years' he kept reminding himself, only seven years until the boys are old enough to go out on their own. In fact, now it was only six and a half years, so, you know, progress.

Something whooshed past his head and smashed into the window frame, sending shards of a porcelain-type material all over Knuckles. He surveyed his body for damage – only a couple of small cuts. He turned to the source of the flying object to see, Balart – the biggest of his boys – smiling at him, another object in his bottom left hand.

'Baly, no!"

Knuckles knew he didn't have long to act, he crouched down and tried to close the gap between him and his least human-looking child before said object became weapon number two. But Baly knew what his dad was doing and finished his game of 'throw' with a well-aimed pitch into the body part that made Daddy react the funniest.

Tears welled in Knuckles eyes as pain shot up from his crotch. He found an inviting piece of floor and assumed the foetal position until the agony subsided. At least that was the plan, cruelly foiled by Baly, then his other two sons, Fjard and Michael, jumping on his prostrate body as the game moved from throw to bounce.

Wendy Wendy, alerted by the commotion, burst into the room, about a minute too late, and let out a fearful roar that indicated it was game over. The boys scurried from the scene leaving Knuckles in a crumpled, quivering mess before his partner.

"Are you alright, my lover?" wailed Wendy Wendy.

"Huurrrpphpffff."

"That's a relief."

Knuckles lifted his head far enough off the floor that the drool chain connecting his mouth and the ground snapped. "A relief? On what planet does huurrrpphpfff mean I'm fine?"

Wendy Wendy moved in to aid her partner to his feet. "I just thought—"

Knuckles shrugged her away and attempted to reach a standing position under his own steam. It took several attempts each generously peppered with wincing and swearing.

Tears rose in Wendy Wendy's eyes as she watched the father of her children process his inability to raise his boys.

Knuckles assessed the damage by touching every sore part of his body until he winced with pain. He went to speak but was interrupted by something being smashed in the other room. He sighed long and hard until the echoes trailed away. "I'm useless at this."

"You're not useless, lover," said Wendy Wendy, "you just need to show them your male strength."

"That was my male strength. I have no more male strength to chair."

Wendy Wendy corrected Knuckles Gartorgian mispronunciation of 'give', which didn't help the situation. He screamed in frustration.

"I can help you. Others can help, other men can help make them strong."

"What kind of father can't make his boys strong? I will be useless to them. And what about when they start hunting? How can I help then? Or schooling? I can barely speak the lineage."

Wendy Wendy resisted the urge to correct him again. Instead she cried. She grabbed her lover tight, wailed and prayed to the gods of Gartogia for the strength to have true insight. Her mind raced, running over all the troubles they faced – inter-species romance, societal rejection, her partner's inadequacies in her world, her children – each different – each neither gartogian nor human. She thought and squeezed.

"Ow," came the muffled voice of her partner, lost in her furry arms.

She released Knuckles from her grip and looked him up and down as he removed a few random hairs from between his lips. Then it came to her. The gods had allowed her true insight, but it would hurt all of them.

"Wait here," said Wendy Wendy before bounding out of the room.

Knuckles removed his taugtaug – a Gartogian robe made of animal pelt – and surveyed the damage to his ribs. Bruising had surfaced in some places already, it just made the pain return again.

Wendy Wendy appeared at the door, sheepishly leaning on the frame.

Knuckles eyed her up and down. "What?"

"Remember the things I do are out of love."

"What? What things?"

Wendy Wendy stepped through the door, she was carrying a bag which she offered to Knuckles. He inspected the contents, then swore.

"My things! You had my things all this time?" Knuckles rifled through the clothes to successfully search out the communicator. Uncontrolled emotion overcame him, he slumped to the floor and half laughed, half cried while cradling the device. "Why?"

Wendy Wendy ignored the question. "That thing has been becoming more and more active recently."

But Knuckles was already accessing the messages, playing the latest from Cooper's UT avatar. As the hologram played out on the floor in front of him, it didn't take long to realise his friends were in trouble. He turned and glared at Wendy Wendy. She started to cry and blurt out, 'I'm sorry' between bass-toned wails.

Knuckles eyed her for a while, a mix of rage and guilt at future freedom rose within him. "I'm leaving," he said as he made his way past her to the door.

He stomped to their bedroom to pack a few possessions, Wendy Wendy followed, pleading with him to talk. He ignored her as he found a light travel pack and shoved everything he considered worth hanging onto inside.

When he was finished, he looked at her again. "Why?"

"Because I knew you wouldn't stay."

"You lied to me, I can't stay now."

Wendy Wendy let out a roar and buried her head in her top set of hands. Behind her, Knuckles saw three sets of eyes peering in from the doorframe.

Knuckles knew this was the moment, he let out a long, slow breath. "Boys, come here, Daddy has something to stay."

"Say," Wendy Wendy corrected though tears.

*

"OK, it looks like we have a couple of minutes before all hell breaks loose," said Cooper. "I reckon we're best to get ourselves side of stage and try to get their attention."

Rhiannon nodded her agreement as Cooper opened the door and they began to negotiate the underbelly of corridors to the stage.

"If we can position ourselves on either side of the stage, surely one of us can catch their eye. You got the extra pistols for Pete and Ginge? Can you flick a couple my way?"

"Yeah," said Rhiannon as she reached into her satchel.

It was then she noticed some unexpected flesh coming from Cooper's pants. "Erm, Cooper."

"Alright, I'm pretty sure these steps lead us to the side stage area," said Cooper over the noise of the enthralled crowd.

"Cooper, your pants."

"Not the time, Rhi."

"Your ass is hanging out of them. Oh my god, so's mine."

Cooper stopped on the top step. "What?"

"You have cut-outs where your pockets should be, it's just… ass."

Cooper touched the area to confirm what his girlfriend had said. "I'm gonna kill him."

"We need to change."

"No time, we need to act."

"There's no way—"

The crowd noise subsided as The Ginge took to the mic again. "To get back to your question, the blue heads, well, they were unbelievable.

"But, if you're looking for something a little, erm," he glanced at Pete who had a look of impending disaster on his face, "Let's just say, out-there, then for my money it was Katralia, she's a Phyllinge. Has anyone checked out the UFO4 Galactic Encyclopedia?"

There was a decent tone of approval from the crowd.

"It's available in all good bookstores and, if you haven't checked it out yet, have a look at the app version, there are plenty of bonus videos and other cool stuff," said Pete, trying to steer the conversation into safer territory.

*

Cooper found a strip of black material folded near the side of the stage. "Here, wrap this around your waist."

"Thanks, Coops. What about you?"

"No time. Just start trying to grab their attention. I'm gonna try to make my way around the other side of the stage. We can't have long."

*

"Yeah, the app's pretty sweet," said The Ginge. "But as anyone who's read anything about the Phyllinge will know, they have the ability to use their…" He looked at the horrified Pete again. "…lady-parts to grab the—"

"Thanks Ginge, I think everyone gets the idea."

"But you didn't let me finish."

"We've got a lot of fans wanting to—"

"Unlike Katralia," said The Ginge as he raised his hand triumphantly for the crowd, who responded with laughter. "Whoo-hooo!"

Pete glared at The Ginge. "Really?"

The MC, a loud-shirt wearer named Alan, leaped at the chance to move the conversation on. "OK, folks, we have time for a few more questions before we move onto the special presentation. Who's next?"

A microphone runner in the crowd was in position next to a UFO4 fan from the front section. She was an A-grade fangirl, paying several hundred dollars for her third-row seat and in full Terenda costume – a figure–hugging red one-piece. What she lacked in Terenda's mind-reading ability she made up for with extra curves. She let out an embarrassed laugh. "Hi guys. I'm Sandra. I just want to say I'm a big fan," she said unnecessarily.

*

Cooper arrived at the other side of the stage and started frantically gesturing to Pete and The Ginge. In the wings opposite, he could see Rhi do the same. But the boys were fixated on Sandra's question.

*

"Anyways, I was just wondering. I mean, like, I know there's heaps of rumours on the internet and stuff, but… and not that you two aren't great, but… where are Knuckles and Cooper?" said the fangirl as the MC tried, unsuccessfully, to cut her off. "I know you said it was a scheduling thing but I can't remember the last time either of them has been seen in public."

*

Cooper's communicator zipped into life and displayed images of the gathering force setting up a perimeter around Madison Square Garden. "Dammit." He looked around the backstage area for a prop or something he could use to grab the boys' attention.

*

"Thanks Sandra," said the MC as he looked at Pete and The Ginge for someone prepared to field the question.

"Rumours?" said Pete. "On the internet? Must be serious."

The crowd laughed.

"The truth is they're just taking a break. It's been pretty crazy in the last few months, but they're doing well and I'm sure you'll be seeing more of them in the very near future."

Sandra giggled nervously. "Oh, OK," she said unconvincingly.

"We still catch up all the time," lied Pete.

"OK, next question," said the MC.

*

Cooper was waving a large red flag he had found from side-to-side, with no effect. He searched for something more eye-catching.

*

The second microphone runner was positioned in readiness next to a large, goatee wearing man in his late forties wearing a UFO4 for President t-shirt. "Hi, my name's

Marcus, big fan. Firstly to The Ginge, you're the man!" The crowd cheered and The Ginge thanked them with a royal wave. "Secondly, Pete, what have you got to say about the way the US government has treated you and the general public since your return? Do you think they fear your position? And what course of action can you guys take?"

Pete cleared his throat. "Thanks Marcus. Yeah, to be honest I'm not sure what to think. We actually want to work with them and, as far as we can see, there's no real crossover between Earth affairs and the galactic stuff, but they've been…"

*

Cooper could see a commotion at the back of the theatre. He put down the rope, which was proving even less useful than the flag as an attention seeker. Now was the time to act. He stepped onto stage. It took a few seconds for the audience to realise a third member of the UFO4 had arrived, but as the collective became aware of his presence an ear-crushing cheer ripped through the venue. He crossed to join the stunned Pete and Ginge, shuffling sideways, ensuring his newly discovered trouser air-conditioning system remained obscured from view at all times.

He grabbed the microphone from a dumbfounded Pete and urged the crowd to quieten down. "OK, people, thank you."

Cooper's words acted as prompt for another round of epic applause. He took the opportunity to pass a hand-gun and arm guard the other members of the UFO4 behind him. Pete and Ginge's expression changed several times as they were greeted with an unexpected combination of firepower and ass cheek. "Careful, it's loaded," Cooper whispered.

"I hope you mean the gun," said The Ginge.

"A – What are you doing here?" said Pete, "And B – Why do you look like the sixth member of the village people?"

"A – We're about to get arrested. B – Gandalf malfunction," said Cooper as he gestured to the crowd to quieten down.

"What?" said Pete.

"They're surrounding the Garden as we speak," said Cooper as he continued to use hand gestures to quieten the crowd.

"What the—," said The Ginge.

"Shut up and listen. I'm gonna create a diversion with the crowd, then we'll sneak off backstage, meet up with Rhi and get Gandalf to zap us back to my place."

Cooper looked out to the audience, now ready to hear him speak. "Thank you. As you can see Sandra, I'm right here. Marcus, you're on the money, my friend, the government is tracking us down, in fact we're about to get raided. Here. Any moment now."

There was some scattered laughter in the crowd. "That's not a joke, turn around, you'll see."

The crowd complied then started booing as from the back of the venue, the SWAT team moved in. Cooper raised his voice to compensate. "We need your help – make noise, panic and run for the exits. Help us and we will be forever grateful."

In an instant the fans obeyed and the Garden bubbled with crazy – screaming, panic, yelling – Cooper urged them on as he, the other boys and the MC, shuffled their way across the stage to Rhiannon. Something whizzed past Cooper's head, and again. "They're shooting, run!"

Everything changed, the noise of firing sent panic through the audience, they ducked, making easy targets of the boys. All of Cooper's modesty was gone as he flashed cheek to the SWAT team and bolted for the side of stage. Bullets raced through the air around him. Cooper raced for Rhiannon's outstretched hands and cover. When he was within a few feet he emptied his lungs and dived, sliding across the floorboards to safety. Behind him, The Ginge, Pete and Alan did the same until they lay in a mangled pile at Rhiannon's feet.

On the bright side, Pete's landing was cushioned; unfortunately it was due to Cooper's exposed cheeks acting as an airbag. Pete quickly tried to retract his face before anyone noticed, but by the time he was upright he was aware all eyes were on him.

The Ginge simply shook his head and said 'eew'.

"Shut up, Ginge."

"If everyone's alright, let's get outta here," said Cooper.

"I think I'm hit," said Alan as he inspected his calf.

The Ginge went in for a second opinion. "Just a scratch, you'll be fine."

"It really hurts."

"Newb."

"Cut it out," said Cooper. "Gandalf, you there?"

"I am everywhere."

"Great, get ready to lift us out. Can you take Alan as well, and patch him up?"

"Indeed. And Knuckles?"

There was a pause as Cooper suspected another flesh pockets moment in progress – "Knuckles? He's AWOL, remember."

"According to my data, he is with you."

Cooper turned in time to see Knuckles teleport to centre stage, clutching his communicator in one hand and a Gartogian weapon in the other.

Time plays funny tricks in those moments, the ones you'll never forget. Cooper remembered screaming "Knuckles, put the pulser down." After that he wasn't sure if a minute passed before a number of bullets pelted the defenseless and disoriented Knuckles, or it happened before the words came out. Either way, the result was the same.

"Knuckles!" yelled Cooper. "Damn it! Gandalf, can you teleport him from there?"

"His communicator is damaged; someone will need to get closer with a functioning device."

Pete peered around the curtain before a volley of bullet fire forced him back. "They're closing in, we don't have time."

"We can't leave him here," said Cooper.

"We can't go out and get him."

"Well, what about we come back and get him later," said Pete.

Everyone looked at Pete. "No way," said Cooper.

"He'll be fine," said Pete in a tone that didn't even convince himself.

Cooper let the change in game plan sink in while he tried to search for a solution. "Got it! Someone give Rhiannon their communicator, her and Alan can teleport back to my place. Pete, Ginge – we're gonna make for Knuckles."

"It's too late, they're nearly on us," said Pete. "Wait, we're doing what?"

"Rhiannon, Alan, go, now. Use Ginge's Communicator. Gandalf, you ready?"

"Ready."

Rhiannon and Alan zwipped out of the venue.

"I got that bit, the them going bit," said Pete, "it's the stepping into the line of fire thing I didn't quite understand."

"Shut-up, Pete," said The Ginge.

"And try to find something white," added Cooper.

"White?" said The Ginge and Pete in unison.

"Yes white, a surrendery tone of white."

"Surrender?" said Pete as several members of the SWAT team rounded the backstage area gun aimed and loaded.

"We surrender," yelled Cooper with his hands above his head.

The Ginge and Pete followed suit and stood, arms outstretched, ready to submit.

"We surrender," repeated Cooper.

***

CHAPTER 4

"Surrender! That was the best plan you could come up with?" said Pete as he tried to find a comfortable sitting position on the unforgivably hard bench.

From the farthest side of the cell Cooper rolled his eyes at the 73rd derogatory mention of his tactics. "This is from the guy who wanted to leave Knuckles dying on the floor in front of us while he saved his own ass."

To say this conversation was followed by a long pause would be underselling it; hours passed, dire prison meals were consumed, around the world countless millions watched videos of cats doing catty things on the internet, laughed, then forwarded them to other catty video watching folk. Then there was another pausey bit.

"Surrender!" said Pete. "Douche."

"You seriously need to shut up."

"I mean, you could not have put us in a worse spot. We have no communicators, no access to the UT, no way of telling the fans what's going on, they have us right where they want us and God knows what they're doing to The Ginge right now. That's not even to mention Knuckles, who's probably dying in a hospital somewhere."

"Surrendering wasn't the plan, you dick, it was just… just wait and see."

"I have things to do, you know," said Pete. "We're supposed to be doing a commentary track for the movie's DVD release today. Tomorrow's bigger – National Geographic were going to be negotiating a new channel, for us. A whole freakin' channel. Well, not us but the galactic brand – cultures, technologies, planets, people – which we essentially own since we have access to the UT and can use any moment from anywhere in the galaxy as vision. You think the movie was big, this is gonna be at another level. Then there's the HBO series, expanding the first movie's story to a ten-part, Terenda spin-off—"

"Well done," said Cooper with no veil on the sarcasm, "and if I hadn't been keeping an eye on you, you would've ended up here anyway. I told you something like this could happen. Did you do anything about it? No. Too busy trying to be the next James Cameron."

Pete thought on Cooper's word for a moment. "Pfft. James Cameron wishes he was the next me. Didn't hear you complain about all the money I've made you."

"The money won't mean anything if we let the governments win."

"So I repeat again, why are we in here?"

"Coz while you've been off making movies I've been planning for this."

"Well… good plan. I can't wait to make the UFO4 sequel: Life Behind Bars."

"You're such an idiot."

There was a commotion at the door, then the familiar rhythm of a series of latches and locks being released.

"Stand at the rear of the room, facing the wall, with your hands behind your back. Now," came the instructions from one of the guards in a tone that suggested he took a little too much pleasure from the authority given by his uniform.

Cooper and Pete obeyed, then heard the heavy door ease open, followed by a number of footsteps, then what sounded like a sack of potatoes being dropped to the floor. It was followed by some moans from The Ginge, the sound of the door closing and the rhythm of the locks again.

Cooper turned to see The Ginge looking pale and bruised. "Jesus, are you OK?"

The Ginge made an unintelligible grunt as he moved to a sitting position on the cold cement floor.

Pete moved in to inspect The Ginge's injuries. "What the hell did they do to you?"

The Ginge looked at the damage to his arms, then padded his hands on his face. "What didn't they do?"

"Did they say who's next?"

"Wow. Just give him a minute, Pete," said Cooper as he brought over The Ginge's now cold meal. "Eat up man; you'll need your strength."

The Ginge shovelled like a starving sumo.

"Seriously though, did they say anything about who they're gonna mangle next?"

"Let him be for five minutes."

"Was it me?"

"Shut up, Pete."

"That's easy for you to say."

"Why? It could be either of us, or Knuckles. Just give it a rest for a moment."

"They're not gonna get Knuckles, he's already been seen to, and they probably think you're the leader still, so—"

"I am."

Pete delivered a mocking snort. "It's me for sure."

"OK, yes, it's probably you. Happy? Now, can you just shut-up for five minutes and let Ginge get his head together?"

"If they so much as touch me, I'll sue the bejesus out of them. Of course, I'd need a lawyer for that, and phase two of your awesome plan seems to involve no lawyers. Surrender, now this? What's next? Maybe we could just maim ourselves and save them all this waiting around and sore fists."

"Oh. My. God. I am going to beat you myself if you don't shut it."

There was another conversational pause as The Ginge ate and moaned; Pete tutted and Cooper did his best to ignore him.

Pete was on the verge of saying 'surrender' out loud again when Rhianna Richardson teleported into the centre of the cell. "Need some help?"

"Rhianna!" beamed Pete.

"Shhhh," said Cooper as he closed the distance to his partner and embraced her to the point of suffocation.

She made a sound that indicated Cooper was a little over-committed to the hug and he released his grip and apologised.

"Did you bring it?" Cooper asked.

"Ahh, I see where you're going, Coops," said Pete, "teleport our way out of here. Brilliant."

"Wrong."

"Wait, what?"

"We're staying here."

"No way. Absolutely no way!"

Rhianna ignored Pete and passed a large folder to Cooper. "It's all there, the message we've sent, their responses and all the interesting research we've collected."

"Get any more good stuff?"

"Oh, I think you're going to love it."

"You're awesome," said Cooper as he started investigating the contents of the folder.

"What's all this?" said Pete.

"Our insurance policy," replied Cooper before embracing Rhianna once again. "You'd better get out of here, Babe, it's not safe."

Rhianna put her hands to Cooper's face and caressed her thumb on his cheek. Tears welled.

"It'll be over soon," said Cooper as he reciprocated the move.

"Get a room," said Pete, "before I gag."

"Oh, I nearly forgot, there's a message bee in the folder, play it when I'm gone. Things just got epic."

She stepped away from Cooper and said, "Ready, Gandalf," before being teleported away as she was mouthing the words 'love you' to Cooper.

"Alright, Coops, what the hell's going on?" said Pete. "And what's a message bee?"

Cooper tried to find a comfortable place on the bench then rifled through the folder. "We are going to be indicted at the UN tomorrow."

"What?" screeched Pete, somehow finishing the word in an incredibly high pitched tone.

"They've been planning it for months. They've been trying to find a way to take our status as Earth's representatives away from us – and maybe some jail time for good measure."

"What?"

"If it all goes ahead as they plan tomorrow, they'll ship us off to the Netherlands to face the International Court of Justice in The Hague."

"What?"

Cooper pulled out the first sheet of paper from the folder. "But I don't think it's going to play out like that."

"What? Why?"

"Because we've requested a pre-indictment hearing with the world's leaders."

Pete rolled his eyes. "Yeah, that's gonna happen."

Cooper tapped the folder, "I'm very confident it will."

***

CHAPTER 5

"So what do we do with this thing?" Pete moved the message bee around from hand to hand, trying to find a clue. "Some sort of button would be handy."

Cooper rolled his eyes, looked over to make sure The Ginge was aware and watching, then said, "Play."

The message bee jumped from Pete's hand and projected a hologramatic image of Bushaii Bartton in front of them. He was tall, yellowish in complexion and angularly ugly to a human eye

"Greetings, Cooper Simpson and fellow champions of the galaxy."

"Pause," said Pete while he stared at Cooper. "Why do you get all the cred?"

"He called us all champions of the galaxy, how much more cred do you want? Play."

The hologram resumed, "It is—"

"Pause. I don't know, maybe a little mention of my name, not just, fellow champion, like I'm your little sidekick."

"Really," said Cooper, "that's your biggest problem right now? Play."

"…and honour—"

"Pause. I mean, I only saved your life, made the movie that exposed Rak Divik and his mates as evil pricks, played it in front of the galactic council and, oh, saved all your asses again with my sweet moves flying UFO in the dogfight near Earth."

"What do you want me to say?" said Cooper. "Well done. Play."

Pete muttered, 'fellow champion', in a variety of disgruntled tones as the message continued.

"Unfortunately I do not make contact today to praise your service or recall past glories of galactic triumph. I bring grave news. It appears, as suspected, that the remnants of the Big Three, now known as The Lore, have not taken the shift in galactic power lightly. War is imminent."

The boys exchanged concerned looks. "While we have numbers on our side, our enemy is far more coordinated and technologically advanced. We need people who know how to fight. We need leadership. We need your help."

The hologramatic image changed from the Fredahn leader to the image of a large spaceship that was very familiar to the boys.

"I assume you remember the Dorsano? It is currently heading to your world with a small crew. The ship is now yours—"

"Pause," said The Ginge, "did he just say what I think he said?"

"Play," said Cooper, not taking his eyes from the message.

"You must fill that ship with the smartest minds, the most talented vehicle operators and the most technologically gifted amongst your kind. My crew will train them, then hand the ship over to your command. There is much to learn and little time."

Cooper, Pete and The Ginge stared, mouths agape as the message continued.

"There is a rendezvous point not far from your solar system; the crew will give you the details. I will meet you there in one of your Earth weeks. Together, we will form part of a larger fighting force. We will train until we are ready to serve alongside each other in war. Until we meet again – stay vigilant."

The message ceased and the room went silent as the three boys bathed in a whirlwind of information.

"Oh. My. God," said The Ginge, eventually, as he slowly rose to his feet.

"The Dorsano," added Cooper. "I have no words."

The pair stared at each other in wonder before the moment was broken by Pete. "No way... there's just no way."

"What do you mean, no way?" said The Ginge.

"I'm not enlisting for war."

"Your planet needs you Pete," said Cooper.

"My planet's perfectly fine," said Pete, "and I don't think my presence is going to make the slightest difference to a galactic war one way or the other.

"Earth's part of this too, you know," said Cooper. "You can't not go."

"Just watch me."

"You're such a pussy, Pete," said The Ginge.

"I'd rather be a live pussy than a dead.... dick. Besides, I've got work to do."

Cooper studied Pete, trying to find the right trigger to sell the mission. "That's exactly what this is."

"What?" said Pete.

"You not enlisting for war, you're... starting production on the next instalment of the award-winning UFO4 series."

Immediately Pete's demeanour changed. His eyes focused, then darted around as he searched all the pros and cons in his mind. "Mmm. Tell me then, this movie, is it a sequel or part two of a trilogy?"

"What's the difference?" said Cooper.

"Well, sequels generally end well for the lead characters; it's not the same with the second part of trilogies."

"OK then, it's a sequel."

Pete thought some more. "Alright, I'm in."

Cooper smiled. "Yes."

Pete and The Ginge joined in.

"The Dorsano, can you believe it?" said The Ginge.

"With our own crew," added Pete.

"In less than a week!" said Cooper as his thoughts turned to the world's governments and the UN. "I'm gonna need you two on this every waking moment from now until we launch."

"But what about The Vague?" said Pete.

"It's The Hague, and I've got that covered. But right now, it looks like we're one week away from leading an interstellar cruiser into battle and we don't even have a crew. We don't have anything. Come up with a plan for what we'll need to get ready before we leave – people, weapons, leadership structure – everything."

Pete and The Ginge looked at each then back at Cooper before simultaneously saluting, "yes, sir." They laughed.

Cooper reciprocated, then looked at Pete. "We've got a lot of ground to cover before we make this... sequel."

"That's right, sequel, totally a sequel and not part two of a trilogy. I have better thing to do than be frozen in carbonite."

"You've got nothing to worry about anyway," said Cooper. "That only happens to heroes, not sidekicks."

*

It was 7pm in the oval office, when President Lloyd finally ushered the last of her white house staff out of the room. She had a good 45 minutes of unwinding ahead of her before the evening's functions began. She poured a drink of water, started some neck stretching exercises and tried to let go of all the problems she faced – domestic and foreign. Just another day at the office, she thought.

Then the most remarkable thing happened. A small, metallic object materialised on the desk in front of her. From nowhere. On the object was a small sticky note, with the words 'play me' handwritten.

Her heart pounded, she jumped from her chair and looked around the room for signs of another person – there were none. "Richard!"

She searched the room again, looking for something out of place – hidden cameras, shifted objects – anything. "Richard!"

Richard Donner burst through to door; it wasn't often his boss yelled out for him. This was urgent. "Yes, Madam President?"

"Was someone in here just now?"

"Erm… No Ma'am," he replied, "… is everything OK?"

"What?" The President was suddenly flushed with self-awareness. "Yes, ignore me. I'll be out in about twenty minutes."

She collected herself. "Have the itinerary for tomorrow's LA trip ready for me when I leave."

"Yes, Ma'am."

As soon as the door closed behind her assistant, she headed straight back to the desk and the magic appearing package. She removed the sticky note and gently touched the strange object. It was cold but otherwise not remarkable. She paused, took a deep breath then said 'play'.

The hologramatic image of Abraham Lincoln appeared on the far side of her desk. He looked himself up and down, the adjusted his waistcoat.

"What… how… who are you?"

"Does my fine country not remember its past, Madam President? I am Abraham Lincoln."

"How?"

"That is not your concern. I am here to remind you of your responsibilities tomorrow."

"I am going to Los Angeles to meet with the Governor and open the new disadvantaged learning centre—"

"You are going to New York to face the UFO4 before you wrongly send them to trial."

"I will do nothing of the sort."

"Oh, I think you will."

There was a drawn out silence as the President studied the hologram without responding.

It was the ghost of president-past who broke the deadlock. "Did you receive your package?"

"I don't know what you're talking about," she lied, "and you have no right being here."

"Ft Lauderdale. 1992."

"Excuse me!"

Behind Abe, footage of President Lloyd's past began playing.

"You were issued an indictment for drink driving, but avoided prosecution. Thanks to your father."

"How did you get this?"

"We see everything."

The President's phone rang – it was Richard, concerned at the raised voices. "Everything's fine" she said before hanging up and returning her attention to the hologram of Abe Lincoln.

"I want you out of here right now," she whispered with venom.

"Your future in that chair depends on you hearing me out."

"I will not," she said at volume before adjusting, "be blackmailed by some idiotic teenagers with some special effects."

"Austin, 1993 to 1997," continued Abe with appropriate video backdrop. "Multiple drug use incidents – marijuana, speed, cocaine – often combined."

"I was at college."

"A lewd encounter with a dorm mate."

"Is this what you are going to show the world? I'd probably gain popularity," she snorted.

"Dallas, 2005; electoral fraud. Washington, 2010 to present; two affairs. Multiple instances of directly lying to your constituents. Manipulating mainstream media to discredit several key rivals both within your party and on the other side of the board."

"I can survive this."

"Austin, 2007; another drink driving incident, another cover-up, this time worse."

The President starting crying. Her darkest moment – the one that followed her wherever she went, whatever her title.

"You hit and killed Emily Stevens, a 24-year-old nurse, on her way home from night shift. Your involvement was covered up. Until now."

"I'm sorry."

*

On the other side of the world, Rhiannon watched the events in the oval office unfold with her new helper Alan, the MC from the Madison Square Garden event. When the President finally agreed to head to New York they high fived each other.

Alan let out a whoo-hoo. "That just leaves Canada and Mexico, the rest are on their way or packing as we speak."

"Even without them we already have the numbers we need," said Rhiannon.

"Rhiannon, you wish to be notified when the Dorsano arrives," said Gandalf.

"Yes, thanks Gandalf."

"The Dorsano has arrived."

"Brilliant! Thanks, Gandalf."

Alan meerkatted himself into the conversation. "Can I have a look?"

Gandalf called up a visual of the ship slowing down into a high orbit over Earth. Two smaller fighter craft flanked her on either side.

Alan let out a noise of gooey excitement that just oozed substantial comic book owner. Rhiannon tried not to laugh as she was equally excited.

"That's. Freakin'. Massive," gushed Alan.

The pair then stared in silence, imagining what it would be like to step foot on a star cruiser.

***

CHAPTER 6

Much had changed for the boys in the last few hours.

First Knuckles had been reunited with the group – like The Ginge, he was carrying injuries but he could walk around under his own steam.

Then they had been taken from their dank, lifeless cell to a holding room at the UN while they awaited their hearing. There was actual furniture to sit on and edible food to ingest – things were looking up.

Even better, their phones and communicators had been returned, surely a sign that their little visits had put the boys in a far better negotiating position.

With their communicators at hand they were able to set about healing Knuckles and The Ginge to their fully efficient best. Ten minutes being healed with the power of the UT would do more than any Earth hospital could manage.

Soon after, the inviting furniture was too much for Knuckles to resist – he snored on one of the couches. Cooper had isolated himself too, picking a quiet corner to rehearse what he'd say to the world's leaders when his moment came. He'd been over his lines so many times that the words were starting to blur together. He was tired, stressed and just wanted it to all be over.

The only conversation to be had was coming from Pete and The Ginge.

"So, hopefully the military side of things is taken care of today – once we get the governments on board we should have access to fighter pilots, special forces, tacticians and whatever else we need. It's the other people we need to concentrate on – caterers, entertainment, relaxation," said The Ginge.

Pete gave The Ginge a suspicious look. "Relaxation?"

"What? Things could get stressful out there."

Pete continued the look.

"What? Just masseuses and stuff, nothing dodgy."

Pete rolled his eyes and shook his head. "What about uniforms? We need to look like a professional galactic operation."

"I'm working on that too – I should have some designs together in the next day or two. What else?"

Pete sighed. "I dunno, there's so much to think about. I do think we need some real star power though."

"What do you mean, star power?"

"Some A-listers. Bring a bit of Hollywood to the table."

"We need fighters, not actors," said The Ginge. "Besides, what actor is gonna actually risk his life out there?"

"Someone who wants to be in the sequel of the highest grossing movie of all time is who. They'll be lining up."

"Whatever, Mr Hollywood."

"Diss me all you want, I'm a big player there right now – we need to use that."

The conversation petered away as Pete started flicking through pictures on his phone. The Ginge studied him. He'd changed so much. He looked different, acted different, sounded different and had an ego the size of the Hollywood sign.

*

Cooper's communicator vibrated, it was Rhiannon.

"Hey Rhi, I'm a bit under the pump time wise here—"

"It's here."

"What's here?"

"The Dorsano."

"Brilliant. Can you do me a huge favour and go and meet the crew, tell them we're putting our team together and we're not far away?"

"Erm… me?"

"Can't think of anyone else."

Rhiannon's heart started racing. She had been helping Cooper since he'd returned. She'd seen things she'd never imagined and she knew that one day travelling to space would be possible but today, without Cooper? "OK. Can I take Alan?"

"Who's Alan?"

"The guy from the Madison Square Garden gig."

Cooper went silent as he filed through his mind, trying to remember.

"The MC, you know, the one with the horrid Hawaiian shirt."

"Oh yes, the porky one."

"Erm, I'm right here," said Alan as he stepped up next to Rhiannon, in view of Cooper on the communicator.

"Sorry, Alan."

He nodded his acceptance of Cooper's apology. "If I may, I have watched every major, and not so major, science fiction show and movie since the 70s. I've collected model craft, I've studied schematics, I know the Enterprise inside and out, The Soluco, Serenity – you name it. I know none of them are real but—"

"Great, you've got the gig."

"Seriously?" said Alan, trying to temper his schoolboy-like excitement with an air of authority and trustworthiness. Without thinking he saluted Cooper. "Yes, sir."

Cooper tried not to laugh. "Can you get in touch with the crew? Get them to send a shuttle craft to transport you. Report back to me when you've got something."

"Will do," said Rhiannon, while Alan saluted once again.

*

"Man, LA's been amazing, so many cool things have happened. I scored an original orc mask, one of the Fast and Furious stunt drivers took me for some drifting spin action in a seriously kickass Ferrari and, my personal favourite, I had lunch with Jessica Alba. That's right, just hangin' with Alba. We were chatting about the role of Terenda in that spin-off film we're planning, she'd be perfect for it," said Pete as he shuffled through his phone to a picture of him and Jessica to show The Ginge.

As annoyed as The Ginge was, teen protocol dictated he had to acknowledge this moment, "Damn! She is super-hot."

"She's heaps nice too."

"I think I might be able to match that, I scored this little sucker," said The Ginge as he pulled a bullet shell from his pocket.

"What is it?"

"A bullet."

"Yeah, no shit, Sherlock. How does that beat Alba?"

"Tom Cruise gave it to me!"

"He did not!"

"True story. He was on Jimmy Fallon the same night as me about a month back. If it was a normal celeb I would've started up a conversation but Tom Cruise is kinda next level, so I just chatted to the other lower class celebs in the green room. Anyways, he comes over to me! Says he's seen our film, and loved it. He thought the bar scene with him from Cocktail was hilarious. Long story short, I started gabbing on about Jack Reacher, which I'd just seen on DVD. You seen it?"

"Erm…. No."

"Tom plays a—"

"Tom? What are you, his mate?"

"Shut up. Anyways, cool film. Tom was a total badass in the fight scenes and there was some serious sniper action. So I raved on about it for a few minutes to him and, a coupla days later, someone hand delivers me this little baby from the movie – Tom had kept a couple of bullets from the set and sent me one."

Pete mustered a level of enthusiasm to The Ginge's story that made clear his reaction was lukewarm at best.

"What?" said The Ginge in his defensive voice. "That's cool."

"It semi-cool."

"It's a live round, delivered from Tom Cruise."

"Dude, I had lunch with Jess-i-ca Al-ba."

"You're a douche. Everyone has a lunch with someone in Hollywood. Besides, you were offering to pay her to be in the film. Tom gave me a gift for no reason."

Cooper popped his head up from his research folder. "Will you two keep it down?"

"Coops, sort this one out for us," said The Ginge. "What's better, lunch with Jessica Alba or a Jack Reacher round from Tom Cruise.

There was a long, long pause while Cooper's face dropped and distorted like he'd licked a lemon… with his eyeball.

"Well?"

"Well, I'm not entirely sure what a jack reach-around is, in fact, and I mean this, I don't want to know. I'd take lunch with Jessica Alba, thanks."

Pete burst out laughing.

"Not jack reach-around, Jack Reacher round!"

"Ewww," snorted Pete between fits of laughter.

Cooper started laughing as the original innocent nature of the comment dawned on him.

The Ginge held up the bullet for all to see. "Jack Reacher, as in the movie. Full stop. Round, as in this bullet."

But it was no good, the other two were now in uncontrollable fits of laughter.

"The thing that makes it funny," said Pete, eventually, "is that Coops actually thought it was plausible that you would go there."

"Seriously. Shut up."

Pete and Cooper reloaded for another round of laughter; The Ginge did his best look like it wasn't affecting him.

Pete, observing The Ginge's 'shut up' request, turned to the art of mime to continue to taunt him.

The Ginge, searched his mind for a clever comeback, and when none dawned on him he threw the bullet at Pete, striking him in the forehead.

"Ouch," said Pete as he pocketed the bullet as punishment.

*

"Presidents, Prime Ministers, Supreme Leaders, Sultans, Taoiseachs and other Heads of Government, thank you for your time on this truly historic occasion.

"As most of you are aware, approximately 15 months ago the four of us were chosen as Earth's representatives to the galactic council. It was a role we fulfilled with dignity, integrity and honesty until the council session concluded. Since our return to Earth, however, there have been strained relations between us, Earth's representatives, and you, the local governing bodies."

Cooper paused for effect.

"We understand the hesitation – the resistance to change. But change is here whether anyone likes it or not. It's pleasing such great strides in diplomacy have been made in the past 48 hours to even allow this meeting to take place. We are repairing the damage and I look forward to working with you all in the future. In fact, we will be calling upon your help sooner than you might think.

"The galaxy is on the verge of war, we have a handful of days to select an elite force of 1500 of the world's finest to man the Dorsano, the ship that will represent Earth in the conflict.

"The timeframe is small, the task enormous and the expectation on those who go will be huge. But we don't live in a fishbowl anymore; we are swimming in the open waters of the galactic ocean.

"Once this meeting is over, we wish to work together with you all to create a task force. Its goal – to assemble the best of the best to join with us as we head back out into the big black.

"In the spirit of open and honest leadership, I will be happy to address any questions from the floor."

Nailed it, thought Cooper as he raised his eyes to see The Ginge, Pete and Knuckles bored out of their minds in the holding room. The final run through of his speech had gone to plan until—

The door burst open, a dozen soldiers, unmarked but dressed in black tactical uniforms piled into the room and aimed some serious looking weaponry at the boys.

Cooper shared confused looks with the others. What was happening?

A man, in his late fifties, entered the room. He walked to the centre and switched his gaze between the four boys, ending on Cooper. "I have a question."

"Who are you?" said Cooper.

"Nice speech," said the man as he narrowed the gap to Cooper. "Inspiring. Naïve, but inspiring."

Cooper's mind ticked over at ridiculous speed. Nothing made sense; he'd got all the leaders where he'd wanted them. He was ready to say his piece. They were on the home stretch, and now this. He didn't know how to respond to the man, so remained silent.

The man studied Cooper for what seemed like an age. "Don't you want to hear my question?"

"Who are you?" Cooper repeated.

"Do you always answer questions with questions?"

"You just did that too," said The Ginge.

Guns swung to line up with The Ginge's face.

"Which is totally OK," he added.

"There's always a funny one," said the man before drawing his attention to Cooper and the communicator on his wrist. "My question is: can you show me the secrets of your technology?"

The man reached in to touch the communicator; Cooper pulled away but was persuaded, with weapons, that resistance was not a good idea. The man took the communicator from Cooper's wrist and put it on his.

He admired the device. "So many possibilities."

"Who are you?"

"Someone most people never meet."

Cooper could only guess where this conversation was headed but he knew enough to be sure of what he said and didn't say.

"Between your chance promotion from very average high school students to Earth's galactic representatives and this thing," said the man before fixing his gaze on the communicator, " you have done something no one else has."

Cooper's head was still spinning; he did not speak despite the man looking at him expectantly for a response.

"You see, Cooper, this world runs on a very specific and refined structure. There are many layers to this structure, many of which you see – business, politics, military, religion – some of which you don't.

"That's where I come in. If those parts you see are the machine that runs the world – the organisation I represent is the oil. We ensure the machine works, whatever the cost.

"So you can see, when four upstart kids start messing with the machine, my organization and I become very concerned.

"I admire what you've done with the world plotting against you. Clever, very clever. But, at the same time, dangerous. Dangerous for my organisation and dangerous for the machine."

"What do you want?" said Cooper.

"Initially we wanted you out of the picture but your actions have persuaded us that compromise is possible. Instead of working against the machine, why don't you become part of it?"

"What? So you would endorse us as Earth leaders?

"That outcome is possible."

"And give us the resources we need to crew the Dorsano?"

"Yes."

Cooper looked at the other boys then back at the man. "And what do you want?"

"This," said the man as he played with Cooper's communicator. "I want to unlock its secrets and understand its power."

Cooper let the information filter through, he looked at the man, his militia and the boys. He needed time. "Can I talk to the boys about it?"

"You can, but I don't see that you have much choice."

***

CHAPTER 7

It had been a dizzying hour for Rhiannon and Alan. A shuttle craft – sleek and silent – landed in front of them at Cooper's remote property, they boarded and launched through Earth's atmosphere into space. Rhiannon let the sensations of fear and exhilaration wash over her in quiet amazement, while Alan offered a running commentary the entire journey. Rhiannon discovered a new talent – the ability to block out the audible register of Alan's voice. His geekgasmic facts on NASA and sci-fi became a background blur in her mind as the knot of fear in her stomach slowly dissipated, only to be replaced by butterflies as she thought of the task ahead.

The shuttle docked with the Dorsano in an automated dance that appeared so normal for the craft yet was so enthralling for its two occupants. A minute later they touched down in the docking bay and the door fffssshhed open. The pair shared an apprehensive look, then exited.

They stood next to the shuttle in the immense and lonely docking area. A countless mishmash of small and medium spacecraft sat in silence. The moment was daunting; chilling even, especially as the shuttle's freezing exterior drew the heat from their bodies. The moment didn't last long as the hologram of an intimidating alien being appeared in front of them.

"Alan, Rhiannon, please excuse this impersonal greeting. Our numbers are few and the tasks ahead are many. Prepare to be transported to the bridge."

*

"Like he said, what choice do we have?" whispered Pete as he and the others huddled in the far corner of the holding room at the UN.

"I don't trust that dude as far as I could throw him," said Knuckles.

"Let's just get out of here without selling our souls," said Cooper. "We've got bigger things to worry about that this guy right now. Agreed?"

The others nodded, the huddle split and they approached the mystery man and his forces.

*

Rhiannon and Alan rematerialised on the Dorsano's bridge teleport platform. Several seconds later Alan opened his eyes, checked all his limbs were in place, squealed like a girl, did a little geek dance and repeated 'like a boss'.

Rhiannon gave him a nudge with her elbow, directing his attention to the greeting party standing right in front of them. Alan immediately adopted a more formal stance and cleared his throat. "Sorry."

"Hi, my name is Rhiannon, this is Alan. Cooper has asked us to greet you and prepare the ship for their arrival." Rhiannon was pleased with the level of authority she weilded.

"Greetings. It is an honour, colleagues of galactic champion Cooper Simpson," said the creature with sharp features and yellow skin who had greeted them in hologramatic form moments earlier. "I am Hensaro Althot, I'm a Fredahn and I will be your primary liaison on the Dorsano."

Beside Hensaro stood a female humanoid, tall, attractive and brimming with pride. "My name Dilania Duritee, I am a Creeaglan from Rovinta, I am the pilot of the Dorsano. It is a great pleasure to hand my ship to the heroic people of Earth."

Behind the two leaders, an exotic array of alien creatures watched on – another Fredahn, shorter than Hensaro, a large and muscular blue creature and three short, smiling smooth skinned beings.

Rhiannon and Alan smiled then exchanged awkward glances. There was so much to take in – new people, new species, new environment, their mission, the apparent importance of humans. They shared a look questioning whether they were in over their heads.

"Rhiannon Richardson and Alan Woodcock, will you join with us as we journey to the bridge?" said Hensaro, as he turned to make the short walk to the ship's operational centre.

"Do you see what he did there?" whispered Alan to Rhiannon. "He used his heads up display to find out our surnames – we never told him. This is so exciting!"

"Excuse me, Hensaro," continued Alan at greater volume, "how do we get access to the Universal Translator so we can access the HUD and other features?"

"A very handy asset to have at your disposal, Alan. Once we have toured the bridge we shall show you to your quarters. From there you can form your connection with the UT and interface with reality on a deeper level."

"Brilliant," replied Alan as he positioned himself next to Dilania. "So... what sort of drive does this baby have?"

"Baby?"

"The ship…. Sorry."

"The Dorsano's primary drive is gravity transfer – seven point, but it can run effectively on three point," said Dilania. "There's also a secondary pulsed warp sphere drive – it's a GF270 running 14 brakcell chambers – it's a relic from the initial ship design before its recent upgrade for war. It only serves as back-up, it's limited on acceleration and manoeuvrability."

That was the moment Alan Woodcock fell in love. "You'll have to show me what she's got some time," he said, reaching for cool but landing in awkward.

Rhiannon saw it all unfolding like a slow motion car crash in front of her, right up until the point where Dilania said, "it would be my pleasure", in what sounded like a somewhat sultry tone.

Alan puffed his chest out, threw his shoulders back and beamed. "I love this ship."

Rhiannon shook her head in disbelief.

*

"Have you reached a decision?" said the man as the boys joined him again.

Cooper looked at his friends. "We feel we can come to an arrangement."

"Excellent."

"But first, what are you hoping to get out of our collaboration and this technology? This will be a technology for everybody when it's established."

"In time, yes." The man studied at the suspicious expressions on the boys' faces. "But you overestimate people. We cannot simply hand out communicators to billions of people; the world would go into meltdown."

"You don't know that. Maybe you underestimate people."

"You really haven't thought about the implications of this, have you? Think about the structure of our world – the powerbase, business, employment, religion – all these things help people find order and meaning in their lives. If this technology were to become widespread, the manufacturing industry would be destroyed overnight – billions would lose their livelihood. You would be destroying countless lives. Yes, it would give people ability and freedom like never before but an occupation is belonging in society. Remove that and most people will be lost. And dangerous. Potentially very dangerous"

The man held the communicator aloft. "You can't just hand this sort of power to the masses. The world, as we know it, would shut down in an instant, with no structure yet carved out to replace it. The rollout of this technology must be strictly managed."

"So, what are you suggesting?" said Cooper.

"In truth, I don't know," said the man. "My organization needs to study what the Universal Translator can do, then put a plan in place to roll it out… for everyone… over time."

"Over what sort of time?"

"I can't answer that yet… but neither can you. I know you position yourselves as 'for the people', but if you really want the best for the people, this needs to be executed properly. It is probably the biggest transition humanity will ever make and we only have one chance to get it right."

Cooper shared a look with the other boys, the collective wisdom from the glances suggested that everyone felt out of their depth with the whole conversation.

The man without a name made some sense. Prior to today, Cooper had been convinced that the people of Earth, when given the technology, would find a way to make it work – in their individual lives and on a global scale. Now he felt unsure, lacking a deeper knowledge and more than a little naïve. At the same time he had no trust for the man – the 'Oil' man, and the organization he supposedly represented.

"So, how do you propose we do this?"

"This ship of yours, what do you need?"

"Erm," said The Ginge as he unfolded a piece of paper from his pocket. "The Dorsano specs recommend a crew of 1500. We're still, erm, sorting it all out but it looks like we'll need about 150 pilots, 300 or so heavy weapons experts, 500 infantry… the rest is made up of command and general support staff, most of which we can gather on our own."

"We can make that happen. We will want some of our people in the command structure."

"Erm…" Cooper tried not to let his suspicion be seen on his face. "Sure… if you can provide a list of candidates we can choose from."

The man looked impressed. "Very well. Let us put the machine into motion, together. I will gather the necessary people to ensure an elite military force we in compiled and that we have buy-in from all the necessary stakeholders."

"And diverse. We want as much of the world represented as possible," said Cooper.

"Indeed," said the Oil Man. "and given our time frame we'd best act now. I suggest we meet back here tomorrow evening, say 19:00 Eastern? We shall select our crew then. The world's leaders are in town so I will ensure all are involved. In the meantime, finalise the numbers you need and send them to me."

"Thank you, mister...?" said Cooper, hoping to have the name blank filled in for him. But his inquiry would go unrewarded; instead an awkward moment started stretching out. Cooper changed tack. "So, do you need anything else from us now?"

Oil Man toyed with the communicator once more. "This."

Not what Cooper wanted to hear. The one asset the boys had was the UT. The communicator gave the man partial access to the UT. He looked around at the armed men; he was hardly in a position to negotiate. Best not fight it. "Sure, I'm not sure how much good it will do you on its own."

The man saw the doubt in Cooper's eyes. "I understand, knowledge is all I seek right now. What say when you return from this mission we plan the rollout of this new technology together?"

The man offered Cooper his hand to seal the deal, Cooper paused then reciprocated.

"Very well. You are free to go. Once you have your final numbers," said the man as he tapped the communicator, "let me know."

"Done," said Cooper.

Oil Man looked at Cooper and the others in turn again before leaving the room, armed entourage in tow. The four boys watched the posse depart, then waited a moment longer before speaking.

"Tell me you don't trust him, Coops," said Knuckles.

"Of course not, but how else are we going to get this done?" Cooper lifted his wrist to his mouth to speak into the communicator before remembering the Oil Man had taken it. "Can someone call Gandalf and get him to bring us home?"

***

CHAPTER 8

There was something about stepping foot on the command deck of a large craft for the first time that had no comparison – a feeling of power and possibility. That feeling doubled when it was a bridge you'd soon be working on and doubled again when it was attached to the largest, most technologically advanced starship humanity has ever operated.

Alan felt it as soon as he entered the expansive, circular space – he could smell, feel and taste it. He breathed in deep, sucking in the power of the room. Instantly it made him walk taller, it made him belong. It made him dare to dream of the small part he could play in human history.

All of which was lost on Rhiannon.

She was full of wonder for the size of the room and the majestic scene it gave as Earth filled half the field of view. Her mouth opened but no words fell out.

Towards the rear of the command deck, but central to the view, sat an array of seats with the captain's chair dominating in prime position – higher than the others and central to everything.

A peculiar thing about the majestic bridge was the draw the captain's chair had on all in the room. It was as if it was made from materials far denser than anything on Earth, causing a gravitational pull to its position. Wherever you were on the bridge you were drawn to the great seat. Maybe it was just the weight of duty borne by those who sat there. Whatever the reason, Alan and Rhiannon found themselves walking towards it as Dilania explained the different workstations.

"...engineering are positioned behind the captain – liaising with the ship's computer system, engine room, weapons functionality, life systems and defensive shields. In front of the captain sits the tacticians – studying enemy tactics, battlefield structure and the status of the Dorsano's fighters. On board weapons and enemy neutralizing fire are controlled from the front of the room."

She positioned herself behind the captain's chair, her chair – for now. "And it's all controlled from here. Captain – to the left pilot and navigation – to the right military command and chief tactical."

"Would you like to take a seat?" said Dilania to Alan.

His heart galloped as he leaned in to caress the arm rest – soft and luxurious. He positioned himself in front of the chair but was filled with a sense of unworthiness.

"Please, sit," urged Dilania,

Alan closed his eyes and leaned back. As the chair enveloped him, its cool material met with his baking skin – calming him. He opened his eyes and exhaled long and slow. Then it hit him – it was something in the elevated height of the seat, or the ability to look over every workstation in the room, or the unforgettable view or maybe there was strange magic in the chair – whatever the case, he felt like he had more control over the world, the universe, than he'd had in his life. He was the captain.

He pictured himself in the middle of battle. He knew enough about the technology aboard the Dorsano to know the minimalist design of the bridge would be transformed in a battle situation – the room would be filled with battle displays, there would be people buzzing everywhere – giving him information and acting on his commands. He breathed in the scent of the room, wanting all his senses to take in the experience. Science fiction, aliens, everything he was teased about for liking as a kid, was real! And, right now, he was sitting in the number one seat in which a human could hope to sit. A wave of emotion hit him and his eyes moistened but he blinked the wetness away before anyone noticed.

That glorious moment only lasted a few seconds before Dilania offered Rhiannon a seat. Alan paused for a brief moment longer before saying 'make it so', laughed, by himself, then exited the throne.

Rhiannon wriggled herself into position on the captain's chair, found a comfortable spot then turned to the others, beaming. "It's… wow!"

"Soon this will be yours to run – Cooper at the helm and you by his side," said Hensaro.

"The Dorsano will be in good hands," added Dilania.

Rhiannon and Alan shared a look between smiles that hoped Dilania's statement would prove to be true.

*

When you've been healed by the Universal Translator, fully healed, you are visited in equal measure by sleep and hunger. Knuckles had made every couch a winner in the sleep contest since his body was repaired. Now his stomach called.

Cooper's lounge room had a creator; it was Knuckles' first port of call after waking from mega-power nap no.3. He ordered enough food to keep a small circus troupe on the go, which promptly materialized in the device, then he returned to the couch and hoed in.

It was just after mid-morning, central Australian time. Pete and The Ginge had taken a break from the spreadsheets and organisational flow charts. The sheer scale of the task ahead was never far from Pete's thoughts. Knuckles' eating techniques were as clumsy as they were loud and Pete let out an irritated groan before deciding food was a better option than frustration. He ordered a burger and chips from the creator, then sat opposite Knuckles on the couch with his bounty.

There was something about Knuckles that had changed. Pete couldn't say exactly what but Knuckles was definitely quieter than he remembered. With everything that'd happened over the last few days Pete realised the four of them hadn't really talked about anything apart from staying alive, or out of custody, or planning for the Dorsano. Now that he actually paid attention it was obvious Knuckles was different. His time away from Earth had changed him, or was it coming back, or was it getting shot by the SWAT team? "So, how're the kids?"

Knuckles grunted dismissively, maintaining his focus on the plate in front of him. "Dude, the kids, they… erm… growing and stuff?"

"Seriously, it's not up for discussion," said Knuckles as he stretched his arms and legs by the couch, still coming to terms with his full range of pain-free movement again.

Pete thought about pressing the issue, but let it go into the ether. Instead he picked up his plate and moved himself back to where The Ginge sat on the floor, surrounded by paperwork. This was no ordinary paperwork; it was the scrambled jostling of the required crew for the Dorsano. Command personnel, pilots, gunners, infantry, medics, tacticians, cooks, entertainers, hairdressers, documenters, reporters, film crew, management, muscle, brains, beauty – everything they could think of. They wanted to draw their team from a diverse a range of industries as possible. Obviously there was a strong need for military, but even then, they didn't want all their fighters to be of military origin. Gamers would be crucial – after all, the boys own gaming experience was a big factor in their survival on Galactica.

But time was the biggest enemy – the project timeline was ridiculous. The logistics for recruiting the best of the best, from a talent pool of seven billion to a functioning team of 1500, including command structures, in less than a week, well, it was impossible.

Fortunately help was on its way.

*

"And rolling," said Jason as one of his assistants snapped the clapper board for take 21.

Cooper's mind sprinted through what he had to say, the tone and inflection Jason wanted and, most importantly the personal reminder not to act like too much of a dick in front of the camera.

"People of Earth," he said, realizing it was very difficult to achieve his last wish when starting a sentence with those words, "we want you."

"We're looking for 1500 of the brightest, most talented amongst you to join us," he finished his segment with the sincere expression he had agreed to with Jason, then internally smiled, his part was nearly over.

Beside him, The Ginge leaned in towards camera. "We need the finest military and support staff on Earth. To join us amongst the stars."

Cooper wasn't sure but he though he saw The Ginge wink at camera out of the corner of his eyes. He tried hard not to roll his eyes.

"…and protect our world from those who wish to do us harm," said Pete, as the dialogue rolled on.

"If you think you've got what it takes to join us on an adventure across the galaxy," said Knuckles as he crossed his arms to push out his biceps, "then we want to hear from you."

"Time's running out – tell us what you've got to offer," said the Ginge.

"And who knows, maybe you'll be joining us amongst the stars," said Cooper.

He stared down the camera as he'd been instructed. He held the gaze for what seemed like a minute before Jason put him out of his misery. "Aaaaannnndddd cut."

The small crew and Jason's entourage broke into applause as nearly two hours of frustration came to an end. The Ginge let out a 'whooo-hoooo!' and Pete offered Knuckles a high five. After Knuckles blanked him, he offered his palm to The Ginge who responded in kind. Pete slipped his shades on and scanned the room to make sure no one noticed Knuckles' fresh-air treatment – success.

Jason sat in observation as the crew moved in to begin packing up the equipment and the boys started to disperse. Observation wasn't a hobby for Jason; it was a way of life. Observation separated Jason from the average person, it separated him from the average reporter and today it would separate him from the people who would still be stuck on this planet next week. He saw a look in Cooper's eyes and knew it was time to play his hand.

He swooped in on his prey, "Coops, you got a moment?"

"What?" replied Cooper, lost in thought. "Oh, yeah."

Jason ushered Cooper to the back of the room, away from the bustling crew and the other boys. "You're not yourself, what's up?"

Cooper was taken aback by the words – or maybe their accuracy. "Just, sorry, so much to do. And that took an hour longer than it should've. That's an hour I don't have."

Jason thought a moment on his next move. "I know it's frustrating, but that hour will save you many, many more. This thing is going to go large. I've got dozens of recruitment people lined up. In two days you'll have a shortlist of candidates – the best of the best from around the world. We talked about this."

"I know," said Cooper. "I don't think Oil Man's gonna be too happy we're doing this – I don't think he's the kind of guy who takes kindly to other people calling the shots."

"It's your ship, not his."

"Ginge and Pete are using their contacts to source people, which is great because I could actually trust the people they recruit, but they're all entertainers and gamers. I'm running through everyone, anyone, who I think could help. It's getting mental"

"Six days, you be launching, you'll have your crew and none of this will matter."

"Or six days and I'll be heading to war with a bunch of people I can't trust who don't have the required skills to help," said Cooper.

"That sounds like last time you went up."

"Good point, well made," said Cooper just as his communicator buzzed into life. It was Rhiannon.

"Hey Coops…," she said in a way that left an obvious empty hole after the words.

Her hologramatic representation glared at him. Cooper searched his mind for what was missing but drew a blank. Instead he settled for an unconvincing, "Hiya."

"So… where are you?"

Then it dawned on him. "I'm supposed to be there, aren't I?"

"Yes! Several hours ago."

Cooper thought about what response could best justify completely forgetting the giant spaceship he was about to captain and his girlfriend who he'd sent on a mission to help him out. The best he could come up with was, "Erm".

"Because you have less than a week to set this ship up for war."

Suddenly he felt the eyes of the room upon him – the boys, Jason, the crew – everyone. "So sorry, just got a couple of loose ends to tie up here then I'm on my way. Promise."

*

Baltwae was a lost world. It drifted through the galaxy without a host star, the result of a catastrophic galactic event millions of years ago. Baltwae was caught in a gravitational slingshot as two stars passed too close to each other. The humble world was sent hurtling from its comfortable orbit around one of those stars to the cold depths of space, forever. This was not good news for the native creatures of the world as their last few weeks of existence were a constant battle to survive epic tidal events, calamitous tectonic movements and furious volcanic activity.

In a time before the Universal Translator and interstellar communications, there was nowhere to run and no one to help. The few who did manage to survive were greeted with a rapidly declining temperature and, before long, death.

The snap-frozen global tomb then drifted unannounced across the Milky Way for millennia before a Fredahn scientific vessel happened upon its existence. A world that travelled off the grid and out of the eye of the Universal Translator was a very rare gift indeed, especially for a once exiled species like the Fredahn. Sure, many worlds were not under the complete gaze of the UT, but to have a world whose existence was not known about was rare indeed.

Rare, but not alone. The Fredahn had made it their business to find such worlds, some of which they'd keep to themselves, others they'd lift the veil on in moments of war – moments like now.

Baltwae was soon to be revealed to the allies of the Fredahn in their fight against the remnants of the Big Three, The Lore. A safe haven in the uncharted black. The planet's core was reignited and energy drawn from the molten magma to power a small colony. It was an outpost from where one of the Fredahn's major contributions to the war effort could be launched.

Bushaii Bartton watched Baltwae approach as his ship pulled into orbit. There was little to see, just a circle of black blocking out the stars behind it. Within the black circle a small cluster of lights signified the only signs of life for a light-year in any direction.

His ship, the Vixor, was the first of many to arrive. From this hidden black oasis in cold, dark nowhere, an attack would launch. Many ships from many species from many worlds, would be represented here, including the humans from Earth aboard the Dorsano.

***

CHAPTER 9

Cooper adjusted his new communicator on his wrist. For some reason it didn't sit in the same comfortable way the old one did. "Gandalf, you there?"

"I am everywhere."

"Knuckles and I are taking the shuttle to the Dorsano, I'll check in again when I get there. Meanwhile, I want a complete background check on Oil Man and monitor everything he tries with his communicator – everything – doesn't matter how small."

"Understood, Cooper."

Cooper had one last look around his room, took a couple of deep breaths to calm himself down then wheeled his suitcase into lounge room where the boys and Jason were waiting for him.

"Alright, Knuckles, let's cruise. Pete, Ginge, we need that list finished by this afternoon. Call me when it's done. Jason can we—"

"What have you got that for," said Knuckles, eyes on the suitcase.

It was only then it dawned on Cooper that he didn't need to take anything with him. With the creator at his disposal he could have any material thing he wanted, whenever he wanted it. He knew this, the last trip away he had left with only the clothes on his back and was fine.

He shot the suitcase a puzzled look, despite the situation not being the suitcase's fault. He put the whole sordid incident down to his primitive early 21st century mindset and shrugged his shoulders. "Bad habits."

"Dumb shit."

"Knuckles!" said Pete, "we've talked about this."

Knuckles let out a frustrated groan, befitting a whining ten year old.

"Talked about what?" said Cooper.

"Swearing."

"I haven't just talked about it, I've been living the swearing dream," said The Ginge.

"Not anymore," said Pete, "we need to cut it out."

"What?" said Cooper and The Ginge in unison.

"Exactly what I said," added Knuckles.

The Ginge and Cooper stared at Pete in disbelief. "Why?"

"Exactly," said Knuckles, reinforcing his previous exactly to the power of two.

"Well, it was Jason's idea really," said Pete, redirecting Knuckles and Ginge's ire accordingly. "Basically, if we drop the swearing back a bit we might be able to get a PG13 rating on the next movie".

"Who gives a shit?" said The Ginge.

"Exactly," added Knuckles, increasing the exactly gravitas.

Pete realized he wasn't getting anywhere; he looked at Jason for support. Jason studied the boys. "Because the marketing folk think if our first movie was PG we would've had anywhere up to another half a billion in box office, that's why."

"Holy shit," said The Ginge in a dreamy kind of way.

"You did it again," said Pete.

"And once you've got the kids in," continued Jason, "the merchandising possibilities take the potential extra earnings into billions.

Cooper resigned himself to a long conversation and moved to the couch. "You do realize with the Universal Translator and the creator we don't actually need money."

"Exactly," said Knuckles, sending the exactlys into exponential orbit.

"True," said Pete, "but if someone's going to give you half a billion, cash, for not swearing, you're going to, right?"

There was a pause in the conversation as everyone thought about the offer. Independently, they were all slightly confused as to whether the choices made sense. Was it a moral dilemma? Did it even matter? Everyone waited for someone else to make the next point.

Until The Ginge switched tack. "Alright then, let's say that's true. And let's say we give a shit about the kids."

"Stop it!" said Pete.

"Why, it doesn't matter what I say here."

"It could do, this conversation could be a crucial part of the next movie."

"What, us? Here now?"

"Maybe."

"Really? People sitting around talking about not swearing? That'd be a pretty shit film, wouldn't it?"

"Ginge!" yelled an increasingly frustrated Pete.

"What?"

Jason stood up, for effect, "I think what Pete is trying to say is we don't know what is or isn't going to a part of the movie yet as we don't know what will happen. So if we can tone down the language all the time, we won't have anything to worry about.

The Ginge crossed his arms and lifted his feet onto the coffee table. "Well, what should I say instead?"

"What do you mean?" said Pete, "Just don't swear."

"I'm not sure that's possible," said The Ginge.

"Of course it's possible."

"No, I really don't think it is. It's how I speak. When I need to emphasise something that little bit extra, I bang in a swear word and presto, fu—"

"Ginge!"

"…king emphasis."

There was another pause in conversation and Cooper made his move. "Look, I'm sure all this is important but Knuckles and I really need to get out of here."

"But if we don't sort it out now it'll be too late," said Pete.

"So we vote on it. Who's pro swearing?" said The Ginge as he raised his hands.

Knuckles raised his hand as well. Cooper saw the way to an easy exit to the conversation and followed suit. "Swearing wins, three votes to one," he said as grabbed the handle on his suitcase and headed to the door.

"What about Jason," said Pete, "he should get a vote on this."

Cooper paused. "Fair enough, three-two, swearers win. C'mon Knuckles, shuttles waiting."

"Wait up, hear me out."

Cooper rolled his eyes and looked at his watch. "Seriously? OK, you've got two minutes."

"Thank you," said Pete. "OK first we have to establish what is a swearword and what isn't. If we use the S H word as the baseline for too sweary we can work back from there to find what's acceptable. Oh, and then there's substitute words, I've got some really exciting ideas around that. For example…"

Pete prattled on. Cooper took a deep breath, thought about all the more important things he had to organize and mentally conjured up a series of swearwords of his own.

*

"Sorry I'm late," said Alan as he entered the observation deck. "Have I missed anything?"

Rhiannon looked at him, she noticed it was the second time in a row she'd seen him with his Hawaiian shirt completely unbuttoned, revealing a white t-shirt and a tuft of greying chest hair. She found his attempt to cool up his image annoying, but put it back to her annoyance at Cooper for still not being on board.

"Not really," she said. "All of the ship reconfiguration plans are on hold until we find out the numbers of personnel we have and their roles. In fact, everything's on hold until we hear back from Coops."

Alan looked at Rhiannon, then at the aliens Hensaro and Dilania. "Still not here?"

Alan never had much success with the opposite sex over the years but he had become quite adept at reading female body language and the look Rhiannon shot him suggested there would be no further talk on the subject. Now was not the time to push his luck and tarnish his image with Dilania.

"Soooooooooo-wah," said Alan as he searched his mind for a good way to change subjects. "What about we continue to skill-up with our ship knowledge?"

He looked around to see approving nods. "Great, Hensaro, why don't you take Rhiannon through the main crew areas? Rhiannon, you can get familiar with the entertainment areas and living quarters. Try to picture how day-to-day life might work for the crew and how we can best take advantage of the space."

"Sounds good," said Rhiannon, her mind elsewhere.

"Great. Dilania, why don't you show me how this baby ticks on a more… intimate level?"

"'This baby'?" said Dilania. "Why do you keep saying that?"

"To sound cool," said Rhiannon.

"Cool?"

Alan felt himself going red as the background colour of his shirt. "C'mon then, no time to waste."

"He's trying to impress yo—"

"She means it's an Earth expression – I'm impressed with the ship," said Alan unconvincingly before overcompensating. "So, Dilania, you still want show me around 'this baby'?"

Rhiannon rolled her eyes.

*

Cooper was annoyed, the two minute resolution to the swearing situation had lasted twenty minutes and in the end they agreed to push the debate back to once they were on board the Dorsano. More time wasted!

Cooper vowed his first order of business once he took charge of the Dorsano would be to make meetings illegal. The punishment for a breach of said law would be to be locked in a room with Pete as he broke down the different categories of swearing – particularly the tricky grey areas between grades C, D and E.

As he was already late, Cooper gave himself another few minutes to go over Pete and Ginge's numbers. The pair begrudgingly handed him a brief case and he sought the privacy of his bedroom to go over the details. Cooper prepared himself for underwhelming but the feelings he experienced never soared to such lofty heights. Inside the briefcase sat one crumpled and folded piece of paper. He picked up the lone sheet, breathed deeply, then unfolded it, being careful not to let the chewing gum stick to the numbers scribbled on the inside.

He ran back into the lounge room. "What the hell is this?"

"Erm, yeah, about that," said Pete, "it might not look like much yet."

"Ya think?"

"But we've got it all in our heads – it's starting to take shape."

"You do realise you'll be presenting this to Oil Man and friends in, like, a few hours, right?"

"Relax Coops, we've got this," said The Ginge.

Cooper looked at the pair. "And when you say, 'got this', I assume you mean, 'we are going to deliver a professional presentation to whichever world leaders are involved'."

"It's all good," said Pete, "sorting through the numbers is the hard part – the presentation's the easy part."

"So, what have you actually got so far?"

*

Cooper's communicator pulsed into life – he was taken aback when he saw it was Oil Man on the other end. He gave Gandalf the signal to display the hologram on the floor in front of him.

Oil Man admired the communicator and Cooper's hologram. "Well this is fascinating, isn't it?"

Cooper saw a smile and relaxed his guard a little. "You get used to it after a while."

"Indeed. How are preparations coming along?"

"Um… yeah… good."

"You seem… hesitant. I trust all will be ready for our meeting this evening."

"Oh, totally," said Cooper in a tone that didn't even convince himself.

"Excellent, I, like the world's leaders, look forward to your recommendations."

"When you say world leaders, who exactly are we presenting to?"

"Well, all of the leaders you… encouraged… to be in New York and a few other interested stakeholders."

"Like."

"The UN – UNOOSA specifically—"

"UNOOSA?"

"United Nations Office for Outer Space Affairs."

"That's a thing?"

"Yes, it's a... thing, as you put it. Plus NASA and the various military interests."

"Oh, OK, so no pressure then."

Oil Man ignored Cooper's sarcasm. "I also have a team working on a ceremony to mark the maiden voyage of Earth's first manned deep space mission."

"Ceremony?"

"Well, once you've wowed us with your plans we need to sell the mission to the world. Hearts and minds, Cooper, hearts and minds."

"Hearts and minds," repeated Cooper as an overwhelming sense of expectation pressed onto him.

***

CHAPTER 10

A fleet of almost fifty medium and large craft gathered near the dwarf star of Yaringa – an assortment of vessels representing various worlds aligned to the New Council.

Yaringa had found itself occupying some crucial real estate when the foundations of a peaceful galaxy had fallen apart at the last council. It represented the front line of the split between the Kenwahla and the Chardrekk.

The Kenwahla – a humanoid species – were in partnership with the Chardrekk – classic greys and Arliff – reptilians in the former council structure. The three species – the oldest and most prolific in the galaxy – were collectively known as the Big Three and had held control of the council for millennia.

Then a new order tore the long-held, but devious, structure apart. A group of species known as the Dihflua faction, led the change for equality, aided and abetted by four teenage humans from a previously insignificant world known as Earth.

Change was swift and substantial. The Big Three split and the majority of grey and reptilian species withdrew from the council. Leader of the council and former Big Three boss, Voim 638, aligned his people – the Kenwahla – with new order. The new allegiance dubbed itself New Council in an attempt to harness the legitimacy their group had earned through political process.

This move left traditional heavyweights Chardrekk and Arliff on the outside. This was not acceptable to the Greys and Reptilians who saw this shift in power as a slight on what they had spent countless centuries creating and protecting. If the power shift was an insult, the ensuing betrayal of the Kenwahla was unforgivable. The Big Three, now only two, changed their name to The Lore and set about seeking to take from those who had taken everything from them – the Kenwahla, some defecting greys and reptilians (or renegades as they were referred), the Fredahn and the four humans from Earth.

Back in the vicinity of Yaringa, the 50 New Council starships were soon outnumbered by a fleet of Lore craft. Surprised and overwhelmed, the battle was as short as it was one-sided. At the centre of the The Lore's vast fleet, a super starship – the Cremmerson –coordinated the attack. The other large ships were dwarfed by its presence. At the helm, Chardrekk captain Isla Enchant, watched the carnage with cold delight. The grey, her ship and her fleet were brutally victorious, the first blow in a new galactic war had been struck.

There was no going back now.

*

Cooper closed the hologram link with Bushaii Bartton and stood for a moment in stunned silence. War. He had known it was coming, that's what he'd been preparing for, but to hear the words, well, that changed everything. War – no longer a concept, now a reality. If he felt underprepared just trying to get things in order for Earth's leaders, that feeling doubled now. They were so far from ready it was scary.

Cooper continued the hunt for Pete and Ginge, which had been interrupted by Bartton's war call. He ran down the stairs and into the second lounge. He was expecting to see the boys working hard at their ship staffing document, instead he was greeted with Jason, deep in conversation on his phone. "Where are they?"

Jason raised his index finger to indicate he was on an important call. Cooper glared at him but he needn't have bothered, Jason had already turned away. Cooper swore, moved to the windows and darted his eyes around the carpark and various sheds he had on the property. No sign of them.

Were they trying to avoid him? They must be. They had to be! Avoiding him because their ship staffing plan was a complete disaster and they were too scared to face him.

Idiots!

He yelled their names as loud as he could and waited. No response.

His thought process was starting to scare him even more; he could feel the sweat starting to bead on his forehead. So worked up he couldn't think straight enough to even work out where to extend his search. He reached into his pocket and found a coin to do the thinking for him. He flipped it – it landed on heads – he would search further outside.

There were only six hours until the presentation. Six hours! Cooper and the boys had lied to him last time he asked for a progress update. His mind had been on a million different things so he had let it go. He should've pushed harder. He should've asked for details. Now a sickly feeling enveloped him. They were walking into the lion's den, totally unprepared. The timeframe, the expectations, everything was set up to see them fail.

Worse still, they weren't answering their communicators – Cooper had no way to find them. Suddenly his communicator buzzed into life, but it wasn't The Ginge, or Pete, it was Rhiannon. He sighed; he couldn't face her right now. He let the communicator buzz away until it stopped.

"Everything alright?" said Jason as he joined Cooper on the back veranda.

"What? No, I need to see Ginge and Pete, like, yesterday. Where are they?"

"No idea. Hey I just got off the phone from the team in Sydney. Have a guess how many applications we've had so far."

"I really need to find out what's going on with the staffing plans."

"They've had over fifteen thousand."

"They're avoiding me, I can tell."

"And the thing is, most of those have come in the last hour. They reckon it could be anything by morning."

Cooper looked at Jason, he realised he hadn't been listening to a single word. "Wait, what?"

"People are flooding us with applications. Quality people, Coops – ex astronauts, former marines, secret service, elite gamers, hackers, artists, leaders – amazing people."

The first bit of good news Cooper had received in a while. He did not want his team consisting of people the Oil Man gave him. The more they could gather by their own means, the better. "That's great, Jase, nice work."

Cooper felt his communicator buzz into life once more; he looked down to see it was Rhiannon again. He called out The Ginge and Pete's names before answering the call.

"Hey Coops… again!" she said in a way that left an obvious empty hole after the words.

Her hologramatic representation tapped its right foot in annoyance.

"I know, I know. Sorry, things are…" said Cooper, without knowing how to finish the sentence.

"You said you'd be here hours ago."

"I know, it's just—"

"If you could see how much needs to be done here, you'd be here already."

Cooper took a few seconds to devise his next move. "OK, so, the thing is the boys are a bit behind with the Dorsano's staffing plans. I was thinking I'd be pretty useless on board until we had that locked dow—"

"No, there's plenty you can do right now."

Again Cooper paused; he wasn't expecting such fiery disagreement. In fact, he'd never seen her like that before. Ever. "Anyways, I was thinking I should stay here to help The Ginge and Pete until the meet—"

"No!"

Cooper started to become aware of the redness in his cheeks. Jason was looking at him, Rhiannon's hologram was glaring at him and now, Knuckles, Pete and The Ginge appeared out of nowhere to stare at him.

"We just really need to finish this—"

"No way!"

Cooper looked from Jason to Knuckles to Pete to The Ginge – their expressions were a varying combination of amusement, sympathy and 'better you than me'. He looked at Rhiannon, her eyes pleaded for help.

"OK, what about I send Knuckles your way and I'll—"

"Knuckles? Seriously?"

Knuckles went to protest, but the look in Rhiannon's eyes convinced him it was a very bad idea.

"C'mon, Coops," said Rhiannon, "I need you."

Everyone stared at him for the magic solution. He had a desire to visit the planet of Serensue – a leisure world in the Garenhaden system his old friend, the late Judrev Oll, had mentioned several times. Apparently it was full of calm waters and extensive beaches and, the best part, the population was capped at ten million. If you found the right place you could go days without seeing another soul, if that was your want. Right now that was Cooper's want.

He turned back in to reality on Earth to find Rhiannon talking to him, while the boys were in conversation with Jason about what to do. But, in that moment's mental vacation, he had an idea.

"Alright!" he said. "Pete, Ginge – where are you at with the plans – no bullshit?"

"Erm," said The Ginge, "we're pretty close."

He stared at both of them. "This can't be half-arsed; we've got to get this by Earth leaders in six hours."

They returned his gaze. "It's seriously close," added Pete.

The Ginge attempted to seal the deal by adding, "trust us."

There was an awkward pause as even The Ginge realised the mere act of uttering the words 'trust us', intrinsically reduced trust levels. Even before Cooper stared at him in that gobsmacked way, he vowed never to say trust me again.

Cooper breathed in deep and let out a large sigh. "The last time I trusted you two, Pete oversaw a massacre of the greatest music legends of all time while Ginge decided ditch us completely to take Earth's galactic relations to Bill Clintonesque levels."

Pete redirected his focus to knocking a rock with his right foot while The Ginge absently played with his communicator.

"Go easy, Coops," said Knuckles, "they've been working their arses off."

Cooper looked at the others, he had been so involved in how much work he had to do that he hadn't realised the others were equally as stressed. He realized now, more than ever, that this was no longer Galactica. It was no longer the four of them doing their own thing, with him trying to hold it all together. It was far bigger than that. Things were epic, worldwide – no, galaxy-wide – epic. It was too big for Cooper to manage on his own, or even with the help of the others. It was time for a change of plan.

"Alright, we need some help. Ginge, Pete once we're done here – I want to see the plan – everything you've got and what you still need to be ready to present later. Jason, get your people to go through those applications, we need, say, three people with both a management and military background to help Ginge and Pete finish off. Actually, get them to give us a shortlist of ten – we'll select the three. They have an hour.

"Oh, and while they're at it, we need some people to take leadership roles in medical, engineering, tactical… all our subdivisions – check with Pete and Ginge for the details. Give us three applicants for each. Pete, Ginge and Jase – you guys can handle these interviews, yeah?"

The response from the group was a general mix of agreement and information overload.

"But isn't it gonna take us six hours to interview all those people?" said The Ginge.

"No, we're spreading the load. In an hour you'll have your three people to help you out with the final stages of the ship plan. It sounds like they'll be the best of the best so if you like them, and they can start straight away – they're hired. Once they're on board it's going to save you massive amounts of time.

Cooper could still see an underwhelmed audience, he continued regardless. "The subdivision leaders aren't as critical for the preparation, so this is only if you've got time. But how impressive would it be if we could present some of the crew at the world leaders meeting?"

"Do we really need to be thinking about that right now?" said Pete.

"Hell yes, we do," said Cooper as he studied the desperate faces of the others. "You guys know as well as I do we're only still in the game here because we have the leaders where we want them. They don't like us, they don't want us, we're just a necessity for them right now."

"Yeah and they can suck a fat one," said Knuckles.

"That," said Cooper. "Or we can put a presentation together that blows them away and makes them realise we aren't their enemy, we're their best chance. We'll give them the ship, the plan and present some of the team.

"Actually, let's make this easier. Knuckles and I can take all the subdivision leaders interviews. Ginge, Pete – you guys concentrate on the presentation. Jason, get that shortlist sorted ASAP."

Unconvincing nods were slowly replaced with hints of belief. Cooper could feel the group in this together once more.

"Yes, sir," said The Ginge as he saluted Cooper, somewhat tongue in cheek.

Pete laughed before following suit. "Yes, sir."

Jason saluted his captain. "Sir, permission to begin coordinating the new appointments?"

Cooper saluted back, "Yes, go. And make them good."

"Yes, sir."

"Ahh, what the hell," said Knuckles as he too saluted Cooper.

"Great. I guess my work on Earth is done – until the leaders' meeting. Knuckles, pack your bags—"

Knuckles cleared his throat in an indicating kind of way.

"Actually," continued Cooper, "don't pack your bags, so old-school. We're heading to the Dorsano."

"I'll prepare the captain's quarters," said Rhiannon through the communicator.

"One more thing," said Cooper. "We're officially at war."

"War?"

***

CHAPTER 11

As the shuttle landed on the Dorsano, Cooper could feel a weight lift from his shoulders, partly because of the lower gravitational level but mostly at being able to leave the complexities and expectations behind him. Life was simpler on the Dorsano, he felt it already.

Knuckles hovered near the exit door waiting for it to open.

"You're keen," said Cooper.

"Hells yeah," said Knuckles. "You've got the captain's quarters, so I figure I've got the pick of any other room."

Cooper laughed; the shuttle door fzzd open and Knuckles disappeared from sight. Cooper stepped out to the docking bay and was leapt upon by Rhiannon; he tasted a tear as he kissed her.

"An enthusiastic greeting," said Dilania, smiling.

"Ah… sorry," said Cooper.

He realized there were over a dozen aliens, and Alan, watching on. He released his embrace, but continued to hold her hand for a few seconds before captain cuties called. He faced the alien ensemble – a quirky mix of species whose only unifying thread was the striking blues and white of their uniform. His heads-up display went into overdrive as his field of vision was filled with the names, occupations, species and home worlds of each of the Dorsano's skeleton crew. Once he'd noted that the humanoid Dilania was the acting captain, he subtly tapped his finger on his thumb to indicate he wanted the HUD to display names only. The UT obliged and his view was simplified.

The crew waited for him to speak, eager fans doe-eyed at his presence. He was aware of his reputation, but he still found it strange. "Dilania, crew of the Dorsano, Earth and its people are eternally grateful for the help you are about to give."

Dilania admired Cooper's poise, his humility and eloquence. All of which confirmed she did, indeed, desire him. "It is we who are grateful, champion of galactic equality and captain of the Dorsano, Cooper Simpson."

The words were gushed with so much intensity and so little of the emotional filters common on Earth that it took Cooper by surprise. It also took Rhiannon by surprise. And it completely took Alan by surprise.

Alan stepped in front of Dilania and shook Cooper's hand. "Welcome aboard, Captain, there is much to show you."

Cooper, who was too scared to make eye contact with either Dilania or Rhiannon, appreciated the distraction. He rattled Alan's hand with vigour and turned his attention to the crew.

Alan followed Cooper's gaze to the other crew members. "How rude, Cooper, this is—"

"I am Dilania Duritee." She noticeably pressed her breasts forward. "I am acting captain and will remain dedicated to your service as long as there is anything you want from me."

Cooper fought valiantly, and successfully, to maintain eye contact. He nodded and moved his attention to the tallest of the crew – one of the Fredahn.

"My name's is Hensaro Althot. I am a Fredahn from Vahoona, I am a military tactician."

To his left, the other Fredahn, slightly more orange in complexion, spoke. "My name is Saratan Dif Yana. I am a Fredahn from Coareef. I am a weapons specialist."

As the introductions continued, Cooper used his HUD as much to gather information as to decipher the English translation of their words. They truly were an odd assortment – most notably the impressive blue-skinned Gluff Hurn stood quiet and powerful, while a trio of small beings also stood out. Led by the one known as Edson Pelk, their angelic faces seemed at odds with the hint of mischief in their large, dark eyes. Throughout the introduction ritual, Cooper couldn't help but feel the ever-present stare of Dilania upon him, while the eyes of Rhiannon and Alan were upon her. Things just got awkward.

*

The phone rang. Anne Martin swore and pealed an eye open to check the time – 2.26am – caller ID not recognised. She swore again and put the phone under her pillow to muffle the noise. A few rings later and it fell silent, but that was a false dawn – it beeped to indicate the caller left no message then shortly after started ringing again.

"What?"

"Erm, is that, Anne Martin?"

"This had better be good."

"So… is that a yes?"

Anne rubbed her eyes as she tried to summon alertness.

"Are you Anne Mart—"

"Yes! Who are you and why the hell are you calling me in the middle of the night?"

"Oh, yeah, sorry about that. My name's Wesley Ellis…"

Anne dropped the phone but could still hear the young man on the other end while she fumbled around for it.

"I am one of the so-called UFO4, although I never liked that name really. Bit lame, I thought. Anyways, you may know me by my nickname, The Ginge, or just plain old Ginge when you get to know me better."

There was a paused at the other end.

"I'm here!" said Anne. "Sorry just dropped the phone."

"Did you apply for a role aboard the Dorsano?"

"Yes, yes I did."

"Excellent, we have reviewed your application – strong military and leadership credentials – and, well, long story short, we want you."

Anne paused. "Wait a minute, Dave, is that you?"

"Who's Dave?"

Anne paused again; it was way too early to take all this in. "Never mind. Are you serious?"

"Um, yeah."

"Are you telling me I'm talking to the actual Ginge and you want me to come and work with you aboard the Dorsano?"

He paused, admiring the use of the term 'the actual Ginge'. "Yes, that's exactly what I'm saying."

There was another pause. "… and you're not Dave?"

The Ginge let out a deep sigh. "Who the hell is Dave? Look, do you want in or not?"

"Absolutely."

"Excellent! When can you start?"

"When do you need me?" said Anne as she opened the calendar app on her phone.

"Does 15 minutes work for you?"

"What?"

"15 minutes? Is that enough time?"

"I… I…"

"Seriously, it's like mega-urgent."

Anne walked into the en suite to examine the state of her hair and face in the mirror. "I guess I coul—"

"Great. Put this communicator on your wrist and when you're ready, press the button. You'll be teleported to where we are."

"What commun—"

The device suddenly appeared next to the sink under the mirror. She examined it, put it on her wrist then examined her features in the mirror again. "Fifteen minutes might be pushing it."

"You look fine, just put some clothes on and press the button."

"Sure," said Anne before she had a realization in the pre-dawn fogginess. "Are you… watching me?"

There was an extended silence on the other end of the line before The Ginge responded. "So, see you in 15 minutes, then?"

"Are you watching me?" repeated Anna as she ripped a towel of the rail and wrapped it around her chest to cover all the main points of potential interest.

There was another pause. "A little bit. But nothing pervy."

"OK… so… can you perhaps, not do that?"

"Sorry, yes, absolutely. I had no idea you'd be naked, by the way. I wasn't…"

There was another awkward pause.

"Are you still there?" said Anne.

"Yes."

"OK, can you not be, so I can get ready?"

"Yes. Totally. So you're still coming then. Brilliant. Great. You won't regret it. See you soon."

"Very well," said Anne.

"OK so I'm hanging up now."

"Go!"

"…and I'm out."

*

"And this is the Captain's quarters" said Dilania, "Your new home."

Cooper entered the room. It was magnificent – spacious and luxurious. "Oh, wow."

"If you ever need me, for anything, please don't hesitate to ask. I'll come any time."

Cooper looked at Rhiannon before responding. "Thank you, Dilania. I appreciate it. Right now, I'd like a few moments to settle in."

"As you wish," said Dilania as she stood to attention, awaiting further instruction.

"In private."

"Sorry, yes, sir. Everyone, to the bridge and let the captain prepare his thoughts."

The group began to fall out, with Alan strategically positioning himself between Dilania and Cooper.

"Oh, Rhiannon," said Cooper, "can I have a word with you?"

Cooper didn't have the heart to look at Dilania but was safe in the knowledge Alan was there to console her if required.

"What the hell was that all about?" demanded Rhiannon as the door closed.

"I have no idea," said Cooper.

"Could she make it any more obvious?"

Cooper didn't know how to answer that question; then again, he wasn't entirely sure it was a question. He decided the best course of action was to avoid the whole talking thing and advance straight to affection, so he moved in to hug and kiss her.

She backed away. "I mean, she knows I'm going out with you right? Like, we're an actual couple."

"Hey, it's all good," said Cooper as he moved in for hug attempt two – successful this time, if a little stiff. "To be fair, her species might not even understand the idea of couples."

Rhiannon pushed him away. "What?"

"What?" defended Cooper. "I'm just saying all species are different – really different actually. I'm just saying that, maybe, her type of humanoid doesn't understand relationships… at least not as we do."

Rhiannon blocked Cooper's attempts to return to embrace. "What are you saying; she's some sort of... super-sexual space slut or something?"

Cooper suddenly felt like he was in the driver's seat of an out of control truck, heading down a steep and windy incline, pumping the brakes with no effect. He tried to think carefully on his words but he wasn't sure if it was going to make a difference anyway. "I have no idea. Besides, you can't say slut. I mean, in general, but especially here. She's just… culturally different."

"Oh my God, you're taking her side?"

Cooper tried the handbrake but the lever detached in his hand. "What? No! I'm not interested anyway. Seriously. Can we talk about the Dorsano? Please?"

Rhiannon looked long and hard at Cooper before bursting into tears. She mumbled something through sobs as he embraced her. They stood in each other's arm for a few minutes without words passing between them. For Cooper, it was the first waking moment he hadn't been thinking about governments or the Dorsano since he could remember. He squeezed her tighter.

*

Cooper felt refreshed as he entered the upper command decks of the Dorsano for the first time in months. By his side, Rhiannon matched his stride and vigour. The main command deck, which connected to the bridge, was a sea of workstations that would look spectacular if the ship was in full battle mode, thought Cooper.

"Captain on deck," announced Edson Pelk, a member of the trio of small aliens.

Cooper sensed the increased activity around him, as well as being reminded he couldn't travel far on this ship without eyes upon his every move. He nodded to Edson and headed down the short, descending path to the bridge.

Upon entering, he found the remaining members of the skeleton crew, as well as Knuckles and Alan.

"Captain on deck," announced Hensaro Althot.

"Really?" said Knuckles, "Am I gonna have to hear that every time he rocks up?"

No one responded.

Cooper toyed with a number of options to acknowledge Hensaro, eventually he settled on a salute. After a moment of visual prompting, the Fredahn returned the gesture.

"Thank you," added Cooper awkwardly.

Knuckles rolled his eyes.

Dilania panthered from her seat at the helm. "Does the captain wish to claim what is his," she said as she made eye gestures to the seat… or her butt… no one was entirely sure. Except Alan.

The Hawaiian-shirted former MC jogged the short amount of ground to where his captain was, again strategically placing his frame between Dilania and Cooper's line of sight. "You've got to try this Cooper, it's amazing," he said between puffs.

Alan ushered Cooper to his rightful seat. The captain looked at his crew's expectant eyes, then at the chair. He felt doubt reach out for him and paused.

"We haven't got all day," said Knuckles.

Cooper took one more look around the room before slowly easing into seat. The chair seemed to reach out to him, snaring him in its soft grip; his heart raced. Something clicked in that union between captain and station – he became the man with the title. He felt it, for the first time, he felt it – he was captain. He looked up at everyone again, a broad smile jumped across his face.

"OK crew, I have less than an hour before I head back to Earth for the leader's meeting. Pete's sent me the latest version of the plans, let's get this ship singing."

***

CHAPTER 12

Cooper heard his name announced over the PA system and made his way onto the stage. He looked up at the crowd but the lights and flurry of camera flashes temporarily blinded him. Instead he switched his focus to the panel of eight seats and took the one where his name was displayed.

There was an uneasy silence in the audience – no applause, no ceremony – in fact, the loudest noise to be heard was the footsteps of Knuckles, who was now making his way to the panel. Cooper could sense the lack of respect as several of the world's most powerful people glared at him from beyond the bright lights.

As the other members of the panel were called onto stage, Cooper reflected on everything that had taken his to this point. The visitation from Rak Divik when he and The Ginge had been in a brawl with Pete and Knuckles, the attack on their space craft en route to Galactica, the message, the attack at The Field, surviving the Earth Party and the Ice Moon and then the battle through the Council Precinct to be heard and change the way the Galaxy was run forever. He thought about the Dorsano – his ship, which turned his thoughts to his lost comrade, friend and mentor Judrev Oll. As he felt the glare of the world's political and military leaders he reminded himself that he didn't need a country to run or a panel of stripes and pins to know he understood leadership and sacrifice – he belonged here. No matter what they thought.

As the last of the panel seats filled up, the low murmur in the crowd died down and all eyes fell upon Cooper. He stood up.

"Leaders of the world, I thank you all for being here today. This meeting, this briefing – is a historic occasion. I don't think anyone here truly appreciates how big it actually is… including us.

"To date, no manned human mission has travelled beyond the moon. In fact, only four humans have been that far… and those four humans have been much, much further." Cooper used a hand gesture to acknowledge the others boys.

"To plan out this mission, in the timeframe we have had, is an astounding feat in itself, the international cooperation required has been unprecedented. May that cooperation continue long into the future.

"Soon this panel, as well as nearly 1500 of the world's most talented, will head to a secret rendezvous point for a month of training before we head to war.

"There are many unknowns ahead, but we do know this mission represents the beginning of a new age of humanity and everyone in this room is a part of that.

"I'll now hand you over to Wesley Ellis and Pete Bates for details of the mission, after which, we will open the floor up for questions. Any concern you have – raise it here. This is your time to be involved."

Cooper paused for a moment waiting for applause, when none came he handed the floor over to Pete and The Ginge and returned to his seat. He looked at his friends as they started to deliver their part of the presentation. Within seconds his sense of dread at how the boys would perform at this moment faded. They spoke with confidence and knowledge and, aided by their new assistants, Anne Martin, Aariv Sinha and Angelo Furlanis, they smashed it. Jason queued the UT into a series of supporting graphics that left even Cooper feeling like he was in safe hands.

His gaze wandered to the crowd – now his eyes had adjusted to the bright lights – he could make out the faces of the world's leaders. He smiled as expressions changed, The Ginge and Pete were involving the leaders, they were winning them over. It was happening, it was really happening.

While Cooper watched on, his thoughts drifted back to memories of Pete slipping in vomit after his first hyperspace travel, and The Ginge threatening aliens with his pulser only to have his pants end up around his knees. He tried not to laugh.

Then came the questions, and they were tough. In most cases the boys had already thought of the situation and had a plan in place. In other instances they came up with a solution on the fly, or resolved to come up with an appropriate resolution before they left.

When they'd finished the applause finally came. Not just applause, a standing ovation. Then the world's leaders joined them on the stage. The boys were presented with gifts from various countries, then a series of photo opportunities followed. Suddenly every president and prime minister wanted to be seen with them. The power shift was complete.

*

"Anyone up for another brewsky?" said Knuckles as he jiggled a six-pack in his hands. "Hells yeah," said Pete.

Knuckles passed Pete his prize then offered out a high five. Pete accepted.

"I'll take some of that action," said The Ginge.

"Think quick," said Knuckles as he threw a bottle Ginge's way.

The Ginge flicked the lid off, took a large mouthful then wiped the excess from his lips and chin. "Whoo-hooo! We. Nailed. It."

"Yo, Coops?" said Knuckles, offering up another bottle.

Cooper nodded in the affirmative, caught the incoming amber missile. "Can you believe that just happened?"

"Absolutely," said The Ginge. "Who was ever going to resist our charm?"

"Good point, well made," said Pete as he took another swig.

"You know, I could have you arrested for this."

The boys looked up to see Oil Man standing at the door of the green room – he was holding a folder but, thankfully no armed protection this time. Cooper shot the other boys a concerned look before returning his focus to the man.

Oil Man smiled. "You do realise we frown on underage drinking in this country."

"But I'm 18," said Knuckles. "In fact we all are… except The Ginge."

The Ginge glared at him.

"And the drinking age in the United States is 21," replied Oil Man.

"Oh…"

"Don't we get, like, diplomatic immunity or something?" said The Ginge.

Oil Man smiled again. "I'm sure that can be arranged."

The Ginge pumped his fist and had another swig.

Knuckles offered Oil Man a beer of his own but he declined before turning his attention to Cooper. "You've done well. Passed with flying colours, in fact."

"You make it sound like a test."

"It was. Do you think we'd let anyone do what you've just done?"

Cooper took a small swig from his bottle. "Guess not."

"You may have played it your own way and brought in some of your own people, but we understand why, you've played it smart," said Oil Man, as he stepped towards Cooper and handed him the leather-bound folder.

"What's this?" said Cooper as he opened it and flicked through the contents.

"Help."

The folder contained about 20 pages – each one a dossier with headshot and career achievements.

"Help?" said Cooper.

"You've got your people, out there the world's military and governments are fighting over their people. Here's our people."

"But—"

"These are the sort of people vital to a mission like this. Trust me."

"But—"

"You say but like this is an option. It is not," said Oil Man as he grabbed the folder and picked a file a random. "May I introduce commander Hank Reynolds? Former fighter pilot, long and distinguished career in the skies and at the Pentagon – he'll be part of your command crew."

"Hang on, just a—"

"Negotiation, Cooper, that is how these things work. That is how anything works. Do you think you'd be here without me and my support?"

"Look, Don, or do you prefer Mr Carlton?"

Oil Man nodded. "Been doing some research, I see. But it changes nothing.

Thursday morning there is a launch ceremony as the world wishes 1500 souls bon voyage. The world's media will be there, your families will be there to say goodbye. Just make sure Hank Reynolds and the others will be there too."

Cooper looked at the Oil Man, AKA Don Carlton. He now knew who he was and what he was capable of. He knew he made a far better ally than he did enemy.

"Yes."

***

PART 2:
TRAINING

CHAPTER 13

They came from 134 countries. They came from all corners of the world (despite the world having very few actual corners), and now they were gathered together for the first time. 1500 of the finest pilots, gunners, medics, nurses, commanders, officers, entertainers, tacticians, technicians and engineers – the best of the best of the best.

Cooper looked out over the field at Cape Canaveral. Most of the Dorsano's fleet of ships had made the trip surface-side to ferry the crew and be part of the ceremony. They lay in formation across the field in front of him. He was still amazed by the sheer scale of what they had put together – and what he'd soon be leading.

The Ginge nudged Cooper in the ribs. "This takes me back."

Cooper looked at the media throng below their stage, the hundreds of thousands of fans and the fleet of spacecraft and personnel that seemed to stretch into forever. "What could this possible take you back to?" said Cooper.

"Remember when we joined the rebellion on –"

"I'm just gonna stop you there, Ginge, is this the bit where you're gonna waste my time mentioning some stupid game we played, like, five years ago or something?"

"No! As a matter of fact… erm… actually, I'm not sure," said The Ginge as he closed his eyes to attempt a deeper focus. "Does the planet Huiwwaa ring a bell."

"If you're talking about the game Conquer Quest then yes, yes it does."

"Ahh. Yeah. Sure… And by game you mean an actual game game."

"Yes!"

"Not some military code for a mission or anything."

Cooper looked at the person who would in all likelihood be his most senior advisor as they went to war. "Please. Stop. Talking."

Cooper turned his attention back to the display of ships in front of them. The two gunships and 18 troop carrying vessels were the largest and most notable. They spread out in a massive V formation across the field, their dark exterior reflecting the dawn Floridian sun. Around them were scattered the attack ships, the larger fighter craft, shuttle craft and rescue recovery vessels. In front of each ship, large or small, stood members of the crew, also in formation. Although the final uniforms had not yet arrived, the flight suits – mostly navy blue, with the UNS Dorsano: Mission 01 insignia – designed by Jason's marketing team – looked, well, perfect.

Cooper looked down at his own uniform and insignia then smiled. He and the boys wore a similar design to the rest of the crew but in azure blue. Cooper initially had wanted all the uniforms to be the same hue but had been convinced that the colour coding would help members of the crew recognize who they were communicating with. While the majority wore the navy blue of the military corpse – pilots, gunners and the like – there were also a smattering of yellows, whites, reds, greens and blacks to signify other disciplines.

To the right of the stage stood the boys' families. Cooper looked over and gave his mum, dad, brother and sister a smile. He had spent the last few hours with them in a strange merging of worlds. He'd shown them through some of the ships and fielded tech questions from his dad and siblings, while convincing his mum he was getting a balanced diet and enough sleep.

Now they beamed with pride, close but no longer touchable – a mirage of glorious normal in the over-the-top majesty unfolding around them.

A shuttle craft approached their location to the cheers of the crowd. It drew to a stop in front of the stage, then lowered to the tarmac as the marching band thundered to a crescendo. The door opened and a bunch of politicians exited, heading towards Cooper and the senior crew. Cooper did his best not to roll his eyes at the sight of their showboating, not a good move now that he was on their good side again.

One of the VIPs, the French president, waved to the crowd and then at the stage. Knuckles waved back with equal enthusiasm, while whispering 'ya dickhead' under his breath.

"Knuckles," snapped Pete.

"Sorry, la dickhead?" Knuckles mocked in a bad French accent.

"Knuckles!"

"What? You saw him at the party last night, thinks he's the man. He's a giant—"

"Don't swear."

"…erm… knob jockey."

"Knuckles!"

"What? What's wrong with knob jockey?"

"Alright you two, enough," said Cooper. "Any chance you can pretend to be professional for, like, the next hour?"

Knuckles mocked him with a facial expression while Pete protested with his 'it wasn't me' look.

Soon the politicians were upon them and the ceremony began. There were speeches, music, cheering, marching, flag unfurling, interviews, live crosses – as lame as it was, Cooper couldn't wipe the smile from his face. While he wasn't sure he could entirely trust the crew, particularly Oil Man's special requests, but while the ceremony lasted, it didn't matter.

They had done it. They had their ship, they had their official acknowledgement and soon the Earth and all its peoples' problems would be a distant memory. He and the boys were heading back to space and Cooper couldn't think of anything better.

As the ceremony came to an end, the Dorsano's senior command team, including the boys, were led from the stage by the world's political leaders to the lone shuttle on the tarmac. As they entered the craft, the Dorsano's other crew members entered their vehicles and a variety of hums and pulse sounds filled the still morning air as the ships readied to depart.

Their shuttle rose spaceward to the sounds of the band and the cheers of the crowd. It was followed seconds later by the two gunships, then the troop carriers and the rest of the craft.

The crowd watched on with wonder as 162 craft glided in a giant circle around the NASA launch site before disappearing at high speed into the blood red dawn sky.

***

CHAPTER 14

Navigator Luca Hoffman swallowed hard as several alert signals beeped to life on his interface. "Um, I'm picking up a signal, sir."

Pete watched from one of the pilot's seats on the heavy gunner ship – the Dorsano's biggest fighter class.

"Well done," he said to the young navigator. Pete turned his attention to Cooper in the other pilot's seat. "That's them, Captain. Looks like they're pulling out of warp."

"Good work. Kill the speed, down the power. Let's see what we're dealing with before we show ourselves," said Cooper.

Pete drew his communicator up to his lips. "Ginge; kill the speed and power. Hold for further instructions."

"Roger," said The Ginge from aboard the other gunner.

Around the two gunships, many dozens of attackers and comets did the same. The pack lay in wait.

Luca held quiet until he realised the eyes of Cooper and Pete were upon him. "Oh, sorry, erm, 15 ships – it looks like mostly fighters with two cargo ships," said Luca.

"Agreed, and there's no sign of any further warp activity," said Pete.

"And they're definitely Grey ships?" said Cooper.

Luca looked at his displays again. "Affirmative."

"Weird. What are they doing this far out of their territory? With almost no protection?"

The Ginge's voice filled the communication channel. "Maybe they're travelling light to avoid making a scene."

"Maybe," agreed Cooper. "But why?"

"I've got a bad feeling about this," said Pete.

"Bad feeling?" said Knuckles from behind his turret. "We outnumber them, like, six to one,"

"More like three to one," said Pete.

"Oh yeah, what's 46 divided by 15 then, loser?"

Pete had his UT do the sum in the background while he concentrated on his navigation. "3.0666 recurring."

"See – heaps more sixes than threes," said Knuckles with the smug sense of victory behind every word.

Pete rolled his eyes then refocused on the task at hand. "OK. Ginge, do you want to flank your squad around the back – travel signal-light and stay off comms. We'll head them off and see if they run or gun."

"Roger that."

Within a few seconds the large fleet of fighters separated into two groups – the main pack staying with Cooper's gunner and a smaller contingent following Ginge's ship. As Pete steered his gunner in signal-light mode, he exaggerated each interaction with the controller. He turned to make sure navigator Luca Hoffman could see each move.

"Get ready," whispered Cooper into his comms. "Even in signal-light they're going to pick up on our presence soon enough. He too, nursed a crew member through his instructions, Daniela Rodriguez absorbing every move the captain made. "Just remember, whatever happens don't do anything until you get my sig—"

The gunner shook as it was hit by a burst of pulse fire, powerful enough to knock some of the crew from their feet and temporarily disable limited nav displays.

"What the?" said Pete. "Where'd that come from?"

Luca reclaimed his position in the navigator's seat as his display flickered back into existence. "Um, um."

"Pete, what's going on?" said Cooper.

Pete glared at Luca. "We're working on—"

Another blast hit Cooper's gunner ship, more powerful than the first.

"I see 'em," said Knuckles as he sent pulse fire back at the enemy craft.

"Knuckles, stop it—"

"Just a couple of fighters," said the goateed one as he let off another series of pulses. He clipped one of the enemies. "Cop that, bitch."

The injured fighter whooshed passed the front of Cooper's gunship, followed a second later by the other. A human handful of the Dorsano's attackers immediately pulled out in pursuit.

"What are you doing?" barked Cooper into his comms. "Back in formation!"

But it was too late, they were gone. Suddenly more pulse fire came in from the other direction, taking out three of the attack craft in front of the gunship.

"No!" screamed Cooper. "For God sake, everyone back now."

But he could barely hear himself over the whoo-hoos coming from Knuckles as he slayed another incoming bogey. The frenzy continued in the main command area as a series of interface displays came to life in front of the pilots and navigator.

"I got one!" said another pilot from one of the attackers.

"All Dorsano ships, back in formation now," said Cooper.

But even as the words exited his mouth another series of attackers peeled away from the main formation chasing the second wave of attackers.

"They've seen us," said Luca.

"What?" said Pete.

"The main group of enemies – they're headed this way."

Pete looked to his captain, "They're headed thi—"

"I heard," said Cooper. "Ginge, you there?"

"Yes," whispered The Ginge.

"Bit late for signal-light, mate. You close enough to that main group to do any damage?"

"Can be soon."

"Good. Fire at will, they're heading for us."

"Roger."

"Erm, we've just lost another three," said Luca as he returned his attention to the display in time to see two more Dorsano attackers crash into each other. "Make that five."

"Shivers," said Pete.

"Attackers, back in formation," screamed Cooper.

"Make that eight," said Luca.

"Attackers!"

But Cooper's last signal was more a call of frustration than an order. He watched another group of fighters left formation to chase an enemy craft. Then enemy fire approached from the larger group, now within range. Attackers and the larger Comet ships scattered everywhere – it was a free for all.

"Knuckles, tell the other gunners to focus everything on that incoming squadron."

"Roger that," said Knuckles before relaying the signal.

Pete looked at Cooper, who buried his head in his hands.

"Ten... hang on... eleven," said Luca.

"Whoops," yelled Knuckles.

Cooper kept his head down as he listened to the confused talk from the Attackers and Comets in the dogfight. Terms like, 'what?', 'I can't see' or 'shit' often preceded another noiseless fireball in the space in front of the gunship.

"Fall back," he repeated helplessly.

"Twenty seven," said Luca.

"Remaining or gone," said Pete.

"Gone."

"Drats."

The Gunship shook again as pulse fire hit them from the front – the incoming enemy were headed straight for the dogfight.

"What do we do?" said Pete.

"We're the biggest ship," said Cooper. "I say we aim straight at them."

"What?"

"Twenty nine."

"They'll move."

Pete searched his mind for an appropriately non-sweary expression of frustration, eventually settling on the disappointing and slightly off-trend 'dang'.

Knuckles blasted pulse after pulse into the rapidly approaching enemy, yelling at his team of gunners to do the same.

"Thirty."

"Their getting close," said Pete.

"Hold your line," said Cooper.

"Like, really close."

"Hold your line!"

"Got one!" said Knuckles. "And another."

"Thirty one."

"Whoops."

"Just tagged a couple here too," said The Ginge through comms.

Then Cooper remembered The Ginge. He remembered his squad, trailing the enemy unit. The enemy unit heading right for them, meaning The Ginge was heading...

"Pete, evasive action! Now!"

"What?"

It was too late. As the remaining enemy craft fanned out and away from Cooper's gunship and the remnants of the dogfight, it revealed The Ginge's gunship and the six craft in support. Cooper screamed, The Ginge screamed - everyone screamed.

Interestingly, everyone also instinctively put their arms in front of their face, presumably to lessen the impact of the collision. Despite this move reducing the injuries by approximately 0.000088% it was not enough to save any of them. Not even Knuckles, with an 88 percent increase in protection, was spared.

*

In the main training facility on the Dorsano, the boys and another 148 members of crew removed their 5S apparatus – there was a palpable sombre vibe from another failed flight simulation. Cooper shook his head as his mind raced through everything that had gone wrong; at how his leadership had failed them.

The Ginge patted him on the shoulder. "We'll get 'em next time."

Cooper didn't respond, instead he tried to muster a positive air before telling the crew. "OK break for lunch. Debrief at 13:00 and... what are you?"

"Group E," said Alan, after consulting his clipboard.

"Let's see... you lot have another sim scheduled for 18:00 tonight."

The group dispersed, leaving Cooper alone. He absently messed around with his hair as he pondered the mistakes he'd just made – mistakes that would cost lives in the real world.

Minutes later The Ginge returned looking for his misplaced communicator only to see the Cooper contemplation. "You alright?"

"What? Yeah, nah, I'm fine."

The Ginge rifled around where he'd detached his game apparatus. "Coops, seriously, who d'ya think you're talking to?"

"I know, it's just... nothing."

"I'm not buying that for a second. Seriously, man, talk to me, I'm all ears."

Cooper looked at his mate, maybe the fact The Ginge was looking elsewhere made it easier to talk. "I don't know. I just thought it'd all come together."

"Uh huh," said The Ginge as he realised he was searching the wrong area for his communicator and moved accordingly.

"I mean, I knew it wouldn't come together perfectly but today has been one disaster after the other."

"Mmm."

"I'm just starting to realise how much more there is to do – how much we hadn't thought about. I know we're going to be training with Bushaii and the Fredahn soon enough, but surely we can present in better shape than this! And it all falls back on me, again. It can just get, a bit, I dunno... there's pressure, it can get... lonely. D'ya know what I mean?"

"Hells yeah," celebrated The Ginge as he held his communicator aloft in a reunion that involved a small jig and a ceremonial restrapping of the device to his wrist. He turned, eventually, to see Cooper looking at him. "What?"

Cooper stared on in stunned silence before the disbelief drew him into a laugh.

The Ginge responded with a laugh of his own, then gave Cooper a little wink to let him know no thanks were required for making him feel better.

"You're one of a kind Ginge."

"Any time, my friend, any time."

"You know what, I think I'll take you up on that," said Cooper as he rose to his feet. "Do me a favour and gather the team together – say 1900. Bring Jase, Alan, Dilania and Hensaro too. It's time to lift our game."

"Whoa, whoa, whoa. The Ginge's party radar is going off—"

"Did you just third person yourself?"

The Ginge ignored him and returned to the mission-critical point he was about to make. "That's launch party time my friend."

"Dude, this is important."

"If you're talking important, you're talking launch party."

"Ginge, I'm serious."

"So am I. And have you seen these people? Wow. Beautiful, athletic, talented."

"We're supposed to be professional—"

"Coops, we're like rock stars to these people. Now's the time to strike, ya know, before they get to know us."

"Dude, seriously, can we—"

Ginge could see his logic was getting lost on Cooper, so he changed tack. "Team bonding – that's what we need, for the mission and stuff. I'll be making sure I bond with as many—"

"Enough!" said Cooper in a raised voice before becoming aware of his tone and reducing to a whisper. "You do realise we're here on a mission, right? Mission first. Always."

"Alright, jeez, relax."

Cooper looked at his friend. Suddenly he became vividly aware of the differences between the two of them. Cooper wore the weight of the world on his shoulders; Ginge was, well, still the same old Ginge. Cooper smiled to himself, maybe he was alone. "Look, I need an hour of your time, tops, but I need you and the others focussed. Then you can go and... bond your brains out."

"Nice! Oh, what are we going to talk about by the way?"

Cooper shook his head and how far down the Ginge's priority list the actual ship and mission were. "That's a conversation that needs to wait for the situation room."

*

The situation room on the upper command decks sat directly below the bridge. It shared the bridge's circular shape; the room was filled with a large circular table, surrounded by 18 chairs. Like the bridge it was a quiet area – any conversations had there would be kept from the eyes of the UT.

In fact, war demanded another level of secrecy from the UT – a standard of privacy not seen in millennia in the galaxy. Quiet areas remained undocumented by the UT, but all activities aboard a vessel or by an individual representing a warring party were only available to be reviewed or monitored by that species. In war, more so than any other time, knowledge and secrecy shaped events like nothing else.

Cooper looked at the faces gathered in front of him – of all the leadership team on the Dorsano these were the ones he totally trusted. The other three boys, his girlfriend, their manager, two aliens and a flabby fanboy called Alan. Everyone else came with baggage – an association with a government or military organisation – some tie that cast suspicion upon their loyalties, at least for now. Trust, Cooper decided, would be something fairly earned.

"Alright, thanks guys," said Cooper. "I won't keep you from the party for too long. But, it has come to my attention that as space-faring warship, we, and there's no easy way to put this, suck."

"The battle sims? I've been looking at the stats, it may not be as bad as you think," said Pete, calling up a hologram display of simulation data and using hand gestures to manipulate through the reams of numbers.

"Really? Because we've outnumbered and out-gunned our opposition on every sim, with every flight squadron group... and do you have the stats on the results?"

Pete rifled through the figures with his hands again. "Erm. There's been 18 sims over the last two days and... just off the top of my head... it'd be about... erm..."

"Eighteen, nil?"

Pete stopped stalling. "Maybe."

"But when you take in to account the conditions," said The Ginge.

"Conditions?" said Cooper. "It's space – it's a void - there are no conditions."

"Just..," said The Ginge before stumbling over his words and sighing in frustration. "Look, Coops, you've got people flying with each other that have never met before – airforce, astronauts, gamers – and from all over the world. They're dealing with the UT and playing with toys ore awesome than they've ever dreamed of. It's gonna take time."

"Well, I'm sure the enemy will take pity on us when they hear we're still learning."

"The thing is," said Ginge just before he realised he couldn't find an actual point and just let his comment pass into forever.

Cooper looked at the others. "Anyone else?"

"The mistakes are those of people keen to impress," said Hensaro. "Behaviours that can be adjusted."

"Communication and coordination will help," added Dilania. "As will your strong leadership from the bridge."

"Plus, everyone's still getting to know each other," said The Ginge. "I've got a feeling things like tonight's party are going to help."

Pete studied The Ginge, "it always comes back to sex with you."

"I never said anything about sex."

"You don't need to, you're so obvious."

"Enough," said Cooper. "Can we just try to focus on getting the flight crew not to kill each other for a minute?"

There was a long silence as they sat around the situation room table like scolded school children.

"I think what we're trying to get out of this," said Rhiannon, "is a plan to improve the training."

"Sorry," said Cooper. "Yeah. Rhiannon's right. Let's work out the extent of the problems and tackle them head on in the morning."

Knuckles kicked his feet up onto the table. "Yeah, good luck with that. It's not just the fight sims where things aren't going well – it's all the way down to the little stuff. Have a look next time you walk down one of the busier corridors – like the main pass into command deck. Some people try to pass you on the left, some on the right. It's crazy."

"So what?" said The Ginge.

"So, we can't even get walking past each other right."

"Probably comes back to what side of the road they drive their cars on in their home country," said Jason.

"What?"

"Those that drive on the left, walk past each other on the left and vice versa. It's just what their used to."

"Well, it's annoying. We should draw arrow markers or something."

"True... and have you seen the cafeteria?' said Pete.

"Coops doesn't hang with the grunts in the cafeteria," said Knuckles.

"Well, if he did, he'd see that most people kinda just hang out with other people from the same country."

"Sometimes it's the pilots hanging together," said Knuckles.

"Or the special forces guys," said The Ginge. "They think their shit don't stink, tossers."

"Ginge!" said Pete. "Language!"

"But even in their little groups you can see the rivalries."

Cooper looked at his advisors in turn. "They're the sort of things I'm talking about. It's like we're not a team, just all these little groups doing our own thing."

"That's normal, though, isn't it?" said Alan. "People gravitate to their own. It's familiar."

Cooper nodded. "I get that. It's just frustrating. When I think back to Galactica it was totally different. I mean, we hung out with the new intakes species all the time. We came from all these different worlds yet we spent plenty of time together. We were excited to be part of this newer bigger thing. Why aren't the crew the same?"

"It will come," said Dilania. "With your strong leadership it will come."

Rhiannon went to say something, but settled on rolling her eyes.

"And with our support," added Alan.

"It is not unusual in my experience. In fact, my people have an expression for this exact phenomenon – without pain or victory," said Hensaro.

Cooper looked at the Fredahn, hoping a deeper understanding of the words would come. They did not.

"Your people, they share an experience and a mission but, for now, it is all a concept. Meaning is only given to that concept when your people share the feelings of pain or the joys of victory."

Cooper pondered the words, then nodded once more in appreciation of the wise counsel. He showed a brave face but beneath the surface he felt more lost than ever. All along he'd thought he was heading to another chapter like Galactica but now he knew that was not meant to be. There was naïve sense of camaraderie and adventure back then – now there was expectation and agenda.

Difference. Leadership. Pain. Victory. They were more than words now – somewhere in their meanings lay a path forward for Cooper, he knew it. He also knew he wouldn't be finding any answers tonight; for now he was just content having a direction.

"See, that's why we need to bond," said The Ginge.

Cooper rolled his eyes, but it soon turned into a smile. As annoyingly one-tracked-mindedly obsessive as The Ginge had become about, well, bonding, Cooper knew the party was probably the best thing for everyone right now.

"OK, Ginge, have it your way. Ladies, you might want to give this guy a wide berth tonight, he's on a mission."

Rhiannon and Dilania smiled before fixing their eyes on Cooper to indicate their desire. Cooper returned the gaze to Rhiannon, despite Dilania's piercing gaze in his periphery.

"Erm, indeed," said Alan, desperate to shift attention away from his captain. "The Ginge, if I recall correctly, has been a strong performer on the party front. Messieurs Bates and Thomas have mixed results in competition against him."

Knuckles felt himself start to redden at memories of he and Wendy Wendy. "Listen here, Chubs. Mention that night again and I'll flatten you."

"I, erm, I, erm, I, erm," said Alan, frozen in a sound loop by fear, despite being twice the age and several kilos heavy than his aggressor.

"I'm sorry, what?" said Knuckles.

"Leave him alone," said Cooper.

"I, erm, I, erm, I, erm."

Knuckles smiled at the fear so visually on display. "C'mon, Alan, use your words."

"Maybe he thinks he could do better," said Pete.

"I, erm, well, no. Certainly not better than The Ginge."

"But better than Pete and I, is that what you're saying?"

"I, erm, I, erm, I, erm."

"I'm pretty sure that was what he was saying," said Pete.

"I, erm—"

"Enough guys," said Cooper. "If things keep escalating at this rate you'll be having a bet. I don't think you need me to remind you how that worked out last time."

***

CHAPTER 15

Knuckles made his way back to the table without spilling a drop from the four pints he carried. He lowered them to the table and fanned three of them out to Cooper, The Ginge and Pete before taking a large swig of the remaining beer.

Pete followed suit, although the drinking pace made his eyes water slightly. He put the glass down, then scratched his forehead with his index finger, strategically drying his eyes with the back of his thumb during the process. "I still can't believe we made that bet."

Cooper shook his head. "Seriously, it happened right in front of me. Like a slow motion car crash... between a Prius and a Smart Car. Tragic. I was thinking surely they're not that dumb... well, I was wrong."

"He had it coming," said Knuckles.

The Ginge took a chug of his beer and wiped the remnants from his mouth. "Anyway, we'll smash him. Plus it'll be a nice lighter moment for Pete's movie."

"Tell me you see what you're doing?" said Cooper looking at his three mates in disbelief. "You can't sucker punch irony like that and not expect a response. Surely you guys, of all people, get that."

Knuckles took another big swig of his beer. "Whatever, he was asking for it. Speaking of which, anyone seen him?"

"Nuh," said The Ginge.

"Probably won't even show," added Knuckles.

"We'll win by default," said Pete.

There was an uncomfortable break in the conversation as each boy thought about the bet, the questionable wisdom in taking it and whether winning by default was anything to celebrate.

Cooper had said his piece, handed the conversation over to the goddess of irony to deal with and turned his attentions to Pete. "How is your movie coming along, anyway?"

"Got some great stuff at the launch – all the ships in formation, launching into space, the massive crowd, approaching the Dorsano, it just looked amazing. The boys from the doco team will start to get some interviews tonight, too. You can all expect a visit from them at some point in the near future. Apart from that, though, it's a bit thin."

"Thin?"

"Just... well, it's tricky to plan when I've got no idea what's going to happen. Last time it seemed so much easier, ya know?"

"Yeah, I think I do," said Cooper as he took another swig form his beer. "Actually, I was thinking about that. You remember when you asked me if this is a sequel or part two of a trilogy?"

"Trust me, if my editing could influence events in any way, this ain't gonna be trilogy: part two."

"You're finding it different, I'm finding it different, so—"

"Wait, you? How?"

"Just... never mind. I'm just thinking why couldn't the first film be, like, a prequel, and this just be the... normal one. That might better reflect how everything's changed."

"Normal one?" said Pete.

"Well, whatever you call it when it's not part two of a trilogy, or the prequel or sequel."

Knuckles drained the rest of his beer and slid the empty glass into the middle of the table. "The quel."

Cooper and Pete looked at each other. "The quel?" they said in unison.

"UFO4: the Quel" pronounced Knuckles

"Knuckles, my friend, your genius has been unfairly understated in my humble opinion," said Cooper before smiling.

"Another round?" said The Ginge as he finished his beer.

The boys nodded in agreement as the redhead went to resupply. Cooper looked around the bar area – although their table was at the back of the room they had a good view of the other crew members going about their unwinding business. It wasn't hard to be conscious of the fact that most eyes were aware of the boys and their movements.

He watched as The Ginge zigzagged his way through the crowd. Some people stared at him, some peeked at him from behind others, looking away if he looked in their direction, some pretended not to notice him at all, only to pay way too much attention when he got close, or say something witty as he moved away. It was as if the crowd had a collective consciousness and it was single-mindedly aware of the boys.

Cooper still found it a weird and awkward sensation, despite having felt it many times before. Since returning to Earth, anytime they were in a crowd – at an appearance or mixing with their fans or the general public – this phenomenon happened. They had termed it meerkatting. Pete and The Ginge seemed to thrive on it, Cooper would notice their personalities rise in intensity one or two notches. Knuckles mostly ignored it, save the odd 'what are you looking at' statement if a male were to stare too long at him. Cooper hated it. It made him self-conscious with every move he made.

Tonight was no exception, although at least they shared the focus of the meerkat phenomenon with an assortment of aliens who mingled in and around the crowd. But they were still oddities to the wider crew population; the boys were a focus for who they were, not their appearance.

Cooper leaned into Pete and Knuckles. "Hey, I think you guys should rethink the bet."

"What? Back out of a bet?" said Knuckles. "Hells no."

"Not back out, can't you reschedule or something?"

Knuckles gave him the 'whatever' stare.

"It's just not tactically smart for us right now. We're supposed to be setting the standard."

"That's the plan," said Pete as he kicked his feet on the table next to Knuckles'.

Cooper rolled his eyes at the unsupportable escalation in Pete's ego. "Really? Look, just don't go making too big a dick of yourselves. We're gonna be on this ship with these people for a long, long time."

"Alright, mum, thanks," said Knuckles.

"Well, can you at least keep the drinking under contr—"

"Alright fellow members of the command deck," said The Ginge as he returned with an armful of drinks. "It's time to get this party started."

Two female crew members appeared behind The Ginge, one carrying cocktails, the other a tray of shots.

"Cooper, Pete, Knuckles – this is Astrid, she's from Norway, and Jemma—"

"Jasmine," corrected Jasmine.

"Sorry, Jasmine, she's from across the ditch."

"What's a dutch?" said Knuckles.

"That was my New Zealand accent," said The Ginge as he turned to Jasmine and winked.

She and Astrid laughed. Pete joined in, Cooper rolled his eyes again.

"You lads want to make some room at the table. Girls, dump your drinks down and I'll rustle up a couple of extra chairs."

Cooper's communicator bleeped into life – it was Rhiannon. "Hey, hon, where are you?"

"We've got a problem."

*

Cooper materialised in the situation room and was greeted by Alan and Rhiannon. Alan – as in, Hawaiian shirted, wardrobe implosion Alan – was the first to catch Cooper's eye. His middle-age spreaded frame was draped in the azure blue of the all-new Dorsano flight crew uniform.

The clothing utterly transformed the comic collector. His podge seemingly reduced to nothing – despite the slim fit of the material that contained it. It was as if his weight had been relocated from his stomach to his chest and arms. He stood as proud as the Milky Way beaming out of the window behind him.

"That looks amazing," said Cooper as he tried to decipher how the subtle lining and texturing remodelled the sci-fi MC. He turned to Rhiannon, covered in a white robe. He studied her expression and started to deduce where the problem might be.

"What have they done?"

"I'm not showing you," said Rhiannon.

"It can't be that bad, can it?"

Rhiannon looked at Alan – an expression that could melt flesh from a face. "Turn around and if you even try to sneak a peek I swear to God..."

She watched as Alan obeyed in silence, then turned her focus back to Cooper before removing her robe.

He looked as her body – the uniform clung to her as if it were spray-painted on – but Cooper knew that wasn't her main concern. It seemed to be cunningly crafted to train the observer's entire focus to the chest region, where two fleshy beacons beamed their presence with all the humility of a sidewalk spruiker. To Cooper, she looked amazing, completely stunning, but perhaps not in a 'for public consumption' kind of way. She crossed her legs uncomfortably, shifted positions and gave him the 'well?' look.

"Wow!"

"My eyes are up here."

Cooper paused to think of the right words to say but could only manage, "Sorry!"

"This was Ginge's doing, right?"

"...or Pete's, but probably Ginge's," said Cooper. "You do look amazing though."

"He's right. Stunning," said Alan as he tried not to leer at her, unsuccessfully.

"Alan!" shouted Rhiannon and Cooper in unison.

Rhiannon reached for her robe, Cooper for his communicator. "Ginge. Situation room. Now."

Alan strutted across to Cooper. "I was just telling her she's just a bit self-conscious right now and she'll get used to it."

"Erm.... I'm still in the room."

The Ginge materialised next to Cooper. "This better be quick, Pete's trying to weasel in on Jasmine as we speak."

He noticed Alan. "There you are, you'd better get downstairs before it's game over." Then he noticed his uniform. "...or maybe not. That looks amazing!"

Alan repositioned himself to what he thought would be the best angle to elicit admiration then nodded with a confident fulfilment the others had never seen before.

The Ginge looked at the robed-up Rhiannon in anticipation. "Ahh going for the big reveal. I like it. Ready when you are."

He offered Cooper a high five moments before Rhiannon groaned in frustration. "What?"

"Stick some wings on me and I'd look like Victoria's Space Secret, that's what!"

"I know, right," said The Ginge before looking at Cooper with his hand still aloft for high five embrace. "C'mon, man, don't leave me hanging."

Rhiannon let out another grunt of annoyance. "He doesn't even see it!"

"It's a bit hard to with your robe on."

"Ginge!" said Cooper. "Not now, mate."

The Ginge slowly lowered his hand as he realised Cooper would leave the high-five incomplete.

Rhiannon marched to the large round central table to where two piles of clothes sat neatly folded. She picked up the significantly larger one. "This is the guys' uniform."

She glared at Ginge then picked up the second pile. In truth, the word pile was somewhat of an overstatement in this circumstance, instead, let's say, she picked up the handkerchief sized garment and unfolded it for all to see. "And this is what the girls get."

"And?"

Rhiannon looked The Ginge up and down. "Don't talk to me."

"What Rhiannon's trying to say is there's a significant amount more flesh showing on the girls' uniform," said Cooper.

The Ginge beamed.

Cooper adopted his 'joke time is over' expression. "...and that's a problem."

"How so?"

"We need everyone to feel comfortable."

"I'm lost."

Rhiannon waved the small piece of material around as a hint.

"Oh that! Of course they're smaller," said The Ginge. "They're made from different materials. They've been individually designed to accentuate both the male and female form."

"And that they do," said Alan before looking at Rhiannon. "You'll get used to it."

"Exactly. Besides, these things are cutting edge. They're inspired by the sci-fi uniforms in popular culture and given a contemporary twist, the designers said it's a great chance to break antiquated social norms and shift paradigms... or something."

Cooper looked at the boys, then at Rhiannon. "Let's just delay the uniforms until tomorrow. We'll come up with a few options, get a few people to test them and get the creator to spit them out when everyone's happy."

"Which would've been a great idea," said Rhiannon, "if they hadn't already been distributed as part of a welcome pack to each and every crew member about half an hour ago."

Everyone looked at the Ginge, only Alan seemed happy.

"Anyone who's in their living quarters will have them already."

At that moment, Dilania appeared in the room and trained her eyes on The Ginge. "Excellent. Knuckles said I would find you here. I just want to tell you what a wonderful job you've done on the uniform. Inspired."

She paraded herself closer the others to draw attention to her outfit but everyone in the room was already a step ahead of her. "This is exactly the sort of uniform an elite flight crew can be proud of. I feel ready to obey any request my captain gives me."

Cooper felt himself going red. He maintained eye contact at all times.

"I will leave you to your discussions and look forward to seeing you at the party shortly. Once again, wonderful work The Ginge." She twirled around then disappeared in a whirlwind of flesh and tight clothing.

Cooper, The Ginge, Rhiannon and Alan stood mouths agape for several seconds not knowing what to say next.

Until, that is, The Ginge spoke. "I don't think I saw her paradigms shift at all."

Rhiannon picked up one of her discarded shoes and threw it at him.

*

"Young Cooper Simpson, to what do I owe this unexpected pleasure," said Mr Johnson.

Cooper looked at the hologramatic likeness of his former science teacher standing in the privacy of his quarters. He looked older and more than a little confused as to why he was on speaking to his former student via hologram. "Thanks for speaking to me at short notice. I'm actually after a big favour."

"Big favour? What favour could the leader of Earth possibly want from a humble former high school teacher? Actually, come to think of it, it probably more appropriate you call me Walter."

Cooper smiled, he always liked Mr Johnson... erm, Walter. "It's, well, a bit of a mission."

"Mission?"

"Yeah, a top-secret mission."

"What could you possibly need me for?" said Walter as removed his spectacles in the way he used to do when emphasising an important point in class.

Cooper was flooded with nostalgia at the move and smiled to himself before responding. "It's complicated."

"I understand complicated."

"...and top secret?"

Mr Johnson played along with a nod. "Of course."

"...and would involve a significant amount of travel."

"Yes." Cooper drew the word out over several seconds.

"Excellent. Would you be willing to start straight away?"

"You do know I've retired."

"So, that's a yes?"

Walter paused. "I wouldn't know what I was saying yes to."

"A top-secret mission at the personal request of Earth's leader."

"If I were to believe what I'm hearing on the radio and seeing on TV, you seem to be surrounded by the world's elite. Why don't you get one of them on your mission?"

"I need someone... outside the system. You in?"

"To join you on the Dorsano and travel to the stars? Absolutely."

"Not exactly."

"Well, what then?"

"That's not a conversation for a communicator. We'll need to meet in person. Can I send a shuttle for you?"

Walter eyes darted as he ran the possibilities through his mind. "A... space shuttle?"

"Yes."

The former science teacher's face slowly collected itself into smile, while his body straightened and shoulders eased back. In one slow move he lifted his right arm to his forehead in salute. "Yes, sir."

*

When Cooper finally returned to the other boys he was carrying two jugs of water and basketful of garlic bread. It was only when he arrived at the table he realised he might need more than that to keep them on the straight and narrow. There was no space on the table to place his additions – jugs, bottles, as well as beer and wine glasses seemed to cover every square inch. Any actual free space had been taken by upturned shot glasses.

He tried to get in close enough to come up with a solution but, from his right, a pang of laughter was followed by one of the female crew – Jasmine – falling sideways on a collision course. With glass in both hands, he was in no position to catch her. Instead, he swiped a significant amount of glassware from the table to minimise the damage of her landing. There was a crash of crew member on table and a shattering broken glass on floor. It was followed by silence then, as she lifted her head and smiled, everyone in the vicinity burst into hysterics. Further afield other groups hooted, laughed and jeered, some yelled 'taxi' – clearly unaware space taxis had not been invented yet, due, in most part, to the lack of demand. Soon the documentary team had their cameras tuned in to the chaos.

At the centre of the core group he saw The Ginge, Pete and Knuckles involved in a drinking game. Not satisfied with just playing it, they were coordinating it, pulling the strings.

He slid the two jugs and the pile of garlic bread into the space had and Jasmine had created on the table then yelled "Who's hungry?"

"Cooper!" yelled Knuckles in the overly friendly way he only managed after drinking... or after trying other, alien, mind-altering substances.

Cooper found it unnerving at best, thoroughly creepy at worst. Then he noticed Pete, straddling a seat backwards, shades on, chewing gum. A visual wave of try-hard wafted passed Cooper and that was before a couple of crew members moved to the side to reveal Pete's right arm. The skin from shoulder to wrist was tattooed – a sleeve that was news to Cooper. He shook his head in disbelief then turned his focus to The Ginge – he had a female crew member sitting on his lap, proudly displaying the new uniform, in fact, they seemed to be intimately discussing how the garb displayed a certain physical attribute.

Cooper wanted to slap the lot of them but it wasn't long before some of their cohorts start screaming his name in tones that forced him to shelve his anger temporarily. A Russian man he'd never met embraced him in a way too enthusiastic way. "You are the biggest legend in the world," he said in accented English through a vault-thick wall of beer vapour.

Cooper leaned backward as best he could until the man slowly released him. He looked at his HUD – Sergey Popov – 26 year old gunner from the Russian military. "Yeah. Thanks," he said as he found adequate separation then turned his attention to the boys.

"What the hell is going on?" he said when within range.

"Bonding!" said The Ginge as he raised his glass aloft then cheersed the air before swigging at his beer.

"Seriously, it's been, what, less than three hours?"

"I know man, where have you been?"

"Oh, nowhere, just, ya know, running the actual ship, on comms with Bushaii Bartton preparing for the whole war thing, oh, and preparing a few words to say to the crew tonight. That's right, then I was consoling Rhiannon and trying to resolve your screw-up with the uniforms—"

"Screw up? Everybody loves them."

"Oh, I'm sure the men love them."

"It's not just the men," said the girl on The Ginge's lap before adding a drunken 'whoo-hoo' for good measure.

"I'm sorry, what's your name?" said Cooper.

"I'm... I... forgot."

Cooper gave The Ginge a look that was a multilayered combination of disbelief, disappointment and disdain. The Ginge clearly missed all three phases of the expression as he responded with a wink before speaking. "This, my friend, is... is... shit, I forgot too!"

Ginge and his friend burst into laughter.

Cooper rolled his eyes, took a mental note of the name displayed on his HUD – Julia Alvanez – and looked around the room. Like before, he was aware of the collective focus on him and the boys, but it wasn't only the meerkatting, there were other looks. Maybe it was Cooper's paranoia, his expectations, his guilt at his failure to deliver to date or the collective output from The Ginge, Pete and Knuckles, but he felt palpable disapproval from certain segments of the crew. Cooper's thoughts never drifted far from Oil Man and the need to prove himself.

He saw Hensaro at the bar and felt drawn to him – or was it repelled from the boys? Either way, he soon found himself in conversation with the Fredahn.

"You look troubled, Captain," said Hensaro.

"What, me?" said Cooper instinctively before realising there were few people on the ship more worthy of confiding in. "It's just... I'm not sure."

"You don't become captain of a starship by being unsure... or without it. Doubt can allow for superior thought processes, if it doesn't dictate decision making."

Cooper tried to wrap his head around the statement, Hensaro was good enough at reading human body language to realise he was struggling.

"Can I offer you some advice, Captain?"

Cooper too, felt his skills at reading Fredahn body language were at the point where he could trust Hensaro like few others. He nodded for his advisor to continue.

"You wear your role of captain as a heavy weight."

Cooper went to reply, but stumbled over his words.

"These people don't need more weight upon them. They each feel pressure to show they belong with you, among Earth's elite – both professionally and socially – they are apprehensive for what lays ahead. Allow them tonight."

"I can see it getting messy though."

"Messy?"

"Look at them. If they keep this up they're gonna be incoherent before the welcome presentation." Cooper laughed in defeat. "By the end of the night... well, it's gonna be a disaster."

"For who?"

Cooper studied Hensaro, trying to guess where he was going. "Them?" he responded warily.

"And how is that your concern?"

"I'm Captain."

"And members of your crew getting too inebriated on a milestone night of celebration before you go to war is going to affect your mission?"

Again Cooper thought on Hensaro's words. "It's just.... I get this vibe like I'm being judged by their behaviour? You know not everyone here is overly thrilled me and the boys are in charge, right?"

Hensaro laughed. "That may be, but who is in charge?"

"I am."

"Exactly."

"Yeah, you don't understand what people – err, human people – can be like," said Cooper, scanning to the fringes of the crowd.

"I understand enough to know that if you lead your people well, they will always call you leader. When you have that, the rest doesn't matter."

Cooper nodded at the council while Hensaro continued. "And sometimes leadership is watching them fall... even if it's into the bottom of a glass."

"What about Ginge, Pete and Knuckles?"

Hensaro and Cooper watched the three boys as they bathed in the limelight and laughed excessively at a joke from one of the females gathered near them. "They don't seem to need any advice at all."

"Yeah, but, look at them. I guarantee you at least one of them will do something monumentally embarrassing tonight."

Hensaro laughed again. "I look forward to it immensely."

Eventually Cooper laughed too. Soon, other members of the crew gravitated towards the pair. They toasted the moment, asked Cooper for anecdotes from Galactica and Hensaro for stories about his world and experiences. They shared tales about their mutual friend Bushaii Bartton. Soon the group grew, as did the stories. Hensaro regaled the crew with tales of humanity's importance to the galaxy, while he cursed the taste of human alcoholic beverages.

Soon Rhiannon joined them. Cooper found himself relaxed for the first time in a while. He was in the moment, a place he visited less and less these days. Tomorrow's problems waited for him in the morning.

He would look back on that moment – that particular moment – as the good old days. It was, in fact, more day than days and even more accurately, hours. And in the morning he would forget most of it, but it would be this moment that he would freeze frame as the good old days.

*

The Ginge bounced up to the stage, beer in one hand fork in the other, to the woots of a large section of the crowd. He chimed the fork against the glass in an attempt to quieten the general din of noise. He took another swig from his beer then struck the fork on the pint again.

"Listen up," yelled Knuckles.

At the same time The Ginge inflicted the third, and fatal, strike on his pint glass, sending glass and the remaining contents to the stage floor in front of him. Now he had their full attention.

He acknowledged the mess and smiled. "Better deal with that before I call the captain on stage."

The Ginge engaged the UT with his mind. He summoned a cleaning cloud, which appeared from a partition in the wall behind him. The crowd oohed and ahhed as the cloud of tiny particles attended to the mess of liquid and glass on the stage.

As the cluster finished its work and returned to the wall with a wave of applause from the captive audience, The Ginge winked towards the doco crew then bowed to further hollers. "I thank you," he added.

"And for my next trick," he joked.

The crowd, especially those at the front, laughed. Cooper shook his head at The Ginge's approval-powered and alcohol-heightened swagger. He was pretty sure he saw the redhead wink at a couple of female crew members near the front. Cooper resisted the urge to roll his eyes once more – The Ginge was enjoying the moment, Cooper told himself.

The Ginge let the applause approval extend a little too long, every now and then stoking the fires of adulation with another pose or expression to ensure the clapping didn't die out. Eventually the relaxed, happy Cooper drew a line in the sand and headed towards the stage.

The Ginge noticed his captain approaching. "Whoa, whoa there, Tiger. We've got to play the intro."

Cooper looked at him in a 'what intro?' kind of way. Behind him, Pete sent some instructions to the UT and seconds later the area above the stage turned into a giant 3D screen. A montage of videos from the boys' time on Galactica was played – images of them in battle on the way to Council Common and fighting off enemy attackers near the Earth. The sound and vision was epic, interspersed with councillors from around the galaxy praising and thanking the humans for their effort in bringing galactic change.

Pete smiled as he watched the crowd get lost in his production. He had painstakingly selected the ideal music and crafted each snippet of footage, each inspiring word about humanity. He had mixed it together to create a piece of work that raised hairs on the backs of necks, that made you want to stand side-by-side with a mate, a stranger, to make a difference. He had made it and now he watched it weave its magic, inspiring his audience for a full five minutes. He looked at Cooper – the captain was surprised and pleased. The presentation finished with a message from Bushaii Bartton. "Warriors from Earth, you have big shoes to fill. I look forward to fighting alongside you shortly."

The music peaked, then faded on a shot of the Dorsano, then the UNS Dorsano Mission badge they wore on their uniforms. As it faded to black, the audience exploded in applause. The boys looked at Cooper's stunned, and impressed, face – eventually he smiled. The Ginge turned his attention back to the crowd, after enjoying the ovation ride for long enough he gave a hand signal to quell the noise. Inspired by the willing crowd, his vocal tone changed to that of a UFC ringside announcer. "Ladies and gentlemen, may I introduce, your captain, Cooper Simpson!"

The crowd erupted again.

Cooper stood centre stage, casting his eyes across the diverse and talented crew they'd assembled. He thought about where they'd all come from, such varying backgrounds and interests, he also though about where they were going and how they'd all make it work together. He looked at the Ginge, Pete and Knuckles.

As the cheering subsided Cooper smiled. "A bit of advice – avoid these three tonight."

The crowd laughed but it soon faded – except for their pocket of drinking buddies. There was an expectant tone in the crowd noise. He wasn't sure whether it was saying impress us or inspire us, but either way, Cooper could feel it.

"This is crazy, right," he said as he looked around the ship and crew. "I mean, this, us, here. It's the start of something. Something big!"

The crowd cheered hard again, letting him know they were hanging on every word. He paused to let them celebrate. With each passing moment he could feel the collective confidence of the group rising, which in turn filled him with an indescribable energy, but one he liked. He was feeling at ease, for the first time in a long time.

"Tuesday... the day after tomorrow," he choked up temporarily. "Tuesday, will be looked back on as one of the biggest days in human history. Tuesday, each and every one of you will be part of the first human-manned starship to set flight into space. I cannot tell you how big this is, it's like, Tuesday humanity leaves the technology age – is that the age we're in? – and enters the galactic age."

The crowd hooted, hollered and applauded again.

"Somewhere out there, there's a war going on. The greys and reptilians have the upper hand, there's a Galaxy that needs us.

"I know some of you feel underprepared. Nothing wrong with that, we're all learning. I just wanted you to know, we are not in this alone. We are heading to a rendezvous point to meet Bushaii Bartton and several other species fighting for the cause. We are going off the grid to learn and train together. We are going there as a starship with potential but we are going to leave elite navigators, gunners, fighters, nurses..."

The crowd cheered with each title.

"...captains..."

Doubly so for that one.

"And we are going to join those other craft in one of the largest military strikes the galaxy has ever seen. The greys aren't even gonna see this one coming."

The crowd cheered again, chanted almost as the tones started to take on a more primitive, ritualistic timbre.

"Time may be our enemy, but I can promise you we are in good hands. And talented people like you will prevail. Who's ready to be a part of the biggest victory humanity has known?"

Roar.

"Who's ready to kick some serious alien ass?"

Roar.

"Alright, raise your glasses. Who's ready for... victory?"

"Victory," screamed the crowd wildly.

***

CHAPTER 16

"Cooper!"

Existence – life – can be defined in black and white terms – alive and dead. Except, that is, when it's on the fringes; like moss, lichen, fungi and hangovers.

"C'mon, Coops, it's your last full day!" said Rhiannon, her pitch seemingly increasing with each word.

With eyes shut, Cooper opened and closed his mouth a few times and decided he was dehydrated. He searched around for the water he kept by his bedside, eventually being passed the beverage by Rhiannon.

He took a large swig, then another. "I feel like shit."

"You look like shit."

That was enough to get Cooper to open his eyes. He searched for a mirror, then for his focus. He rubbed the moisture, sleep and general sense of tired from his eyes and started again with more success. Yep, he looked as good as he felt. He searched his mind for memories to fill in the gaping blanks. He couldn't remember going to bed, or anything that felt like an end of the night activity. He decided to start with his last known thought and fill in the blanks from there.

He took another sip of water. He clearly remembered the music cranking and people dancing, crew crowding around him, saying how inspired they were, the boys and Alan doing their best to win the bet.

Then he remembered Pete showing off his flying skills. Then, ohh... "Is he alright?"

"Pete? Broke his arm and collarbone – the UT's fixed him up though. He put a fair sized hole in the bar and his ego."

Cooper snorted. It was then that he noticed Rhi's attire. "What are you wearing that for?"

She paraded the ground between self-conscious and posing in her official Dorsano uniform.

"I thought you hated it," said Cooper, his baffled eyes trying to focus on hers.

Rhiannon thought about her response. Part of her hated it but part of it knew it made her look stunning. Maybe she didn't hate it, hate's such a strong word, maybe she disliked what it represented, despite looking good in it. But when she thought about, what did it represent exactly? Maybe Ginge was right with what he said about paradigms and stuff. Sure, she could wear anything to help change paradigms, even if it did make her look amazing. Sure the other women – elite, talented women – seemed to find the uniforms empowering, inspiring, even. Look, bottom line, she'd be damned if she was going to have the other girls swanning around looking stunning when she was every bit as worthy of wearing the sky blue.

"I'm a team player."

"But, you sai—"

"Just... team player."

Cooper went to inquire further but her expression suggested it was a bad idea. Instead he looked her up and down. "It looks amazing on you."

"I know."

*

Cooper double-checked himself in the mirror and had Rhiannon do the same. A combination of greasy breakfast and intensely strong and sweet coffee had managed to have him feeling slightly human again. "Gandalf, are you there?"

"I am everywhere."

"Can you put together a news report of events from the war in the last 24 hours? Usual style."

The wall opposite his seat transformed into a giant screen. A pair of TV presenters – based on Jamie and Sarah from The Morning Show – appeared.

"Good morning to you on the eve of a historic day for humanity from the bridge of the Dorsano," said the UT-generated version of Jamie. His voice, movements and mannerisms – all brought to life by the UT – made for a reasonably convincing counterfeit. Good enough, at least, for Cooper not to notice or care.

"Historic indeed," added faux Sarah. "Just over 24 hours from now, we'll see Earth's first foray into galactic war. Those aboard this very ship will head into battle."

Jamie looked around his imagined spot near the captain's chair. "It's a magnificent ship, Sarah, and it's crewed by our best."

"Let's hope they're at their best as reports are coming in that don't look good for the New Council. For more here's Cathy at the news desk. Cathy."

"Pause," said Cooper. "Rhi, you watching this?"

"Uh huh," she responded through a mouth that refused to close.

"Play, Gandalf."

"Thanks Jamie, Sarah. In what has been the biggest movement of the campaign, The Lore forces took out the 87th Armada – one of the biggest New Council fleets. Survivors report an armada of vast proportions and a battle that was as brutal as it was clinical.

"For more on the aftermath of what has been a dark day in the campaign, reporter Bryce Neagle is at the scene.

The view switched to the battle zone – space littered with starship carcasses and the flotsam and jetsam of war. The camera shot started tight on a craft even larger than the Dorsano. At least I would've been had it still been intact, instead a mangled cross section of the inner workings of the behemoth were on display. As Bryce spoke the view pulled further and further away until the virtual reporter was revealed overlooking the carnage from a viewing deck on another craft.

"Even flying amongst the debris of such destruction, it's hard to comprehend the scale of events that took place here in the Jexan System. There were 117 Star Class ships in the 87th Armada – an impressive array of vessels from across the galaxy – most now lay as metallic husks, beyond salvation.

"Survivors report blinding lights and an overwhelming attack that came without notice. It was over in minutes. Such was the speed and ferocity of events, a meaningful response could not be mounted.

As salvage crews warp in from around the galaxy to tend to the injured and count the cost of defeat, attention must surely turn to the apparent one-sided nature of the battle.

"It would appear, yet again, The Lore have the intelligence advantage, an advantage they clinically seized upon here and in dozens of other battles in recent days. For the New Council to turn this war around, it seems it will take more than vehicles and weapons, it will take information.

"Bryce Neagle, Jexan IV."

"Thanks Bryce," said Cathy. "As members of the New Council count the cost, we have a look into what that means for the men and women of Earth who are about to enter the battlefield. Morning Show host Jamie Barnard filed this special report."

"They may have the right stuff, but what exactly—"

"Pause, Gandalf."

The screen displayed Jamie frozen with an awkward expression on his face, decked out in fighter pilot paraphernalia, trying way too hard to look cool and failing dismally to Cooper's eye. Cooper looked at Rhiannon. "That is not good, is it?"

"That's really not good."

"We're not ready for this, are we?"

"I'm scared, Coops."

"Me too."

They sat in silence for a moment, each contemplating the enormity of what they'd seen, of what they may soon face. Cooper looked at Rhiannon, her eyes displayed the fear she spoke about. Her beauty had an odd way of shining in the moments he'd least expect it. Like now, he was hit by a full wave of admiration. Not in the sort of way her new uniform made him feel, it was something different – a look in her eyes or the way she held her head despite what they were dealing with. Poise – that was the word, he decided – poise.

"I'm sorry," he said.

"What for?"

"For getting you involved in all of this."

Rhiannon laughed. "You make it sound like I gave you a choice."

Cooper returned her smile. He moved forward and they kissed.

The moment was broken by the voice of Gandalf. "Cooper, you wished to be informed about Anne and Walter's arrival."

Cooper shared a look with Rhiannon, then sighed. "I did."

"They are here."

"Thanks Gandalf," said Cooper scratching his head. "Rhi, any chance you can get them from the landing dock and kill some time with them for a few minutes? I need to speak to Bushaii.

*

In the hours that followed life slowly started getting back on track aboard the Dorsano. Some rose early to exercise and show the world how responsibly they had behaved the night before. Others joined them, exercising despite what they'd put their bodies through the night before, to show the world how hard-core they were. Some woke next to a living, breathing reminder of their exploits the night before – some of these made hasty exits. Others ordered breakfast from the creator. Some just slept.

As the hours pushed lunch and launch nearer, the scene aboard the ship could not be further from the carefree social celebrations of the night before. This was the most personal of times. There was no escaping the utter truth – they were leaving Earth. What was a distant, dreamy concept, was now an inevitable reality. Beyond that loomed the notion that this mission may not have a return leg, that what was happening out there, amongst the stars, might claim them.

It was a time to speak to loved ones, to write in journals, to blog. Some turned to their religion; some studied the ship and battle, some pampered and preened themselves as they prepared for the Dorsano's big day. Despite the diversity of activities, they were all doing the same thing – grounding themselves in simple pleasures and rituals before life changed forever.

The Ginge slapped water over his face as he prepared to be interviewed by the documentary crew. Pete ran through the footage he'd obtained from the party last night and set about the task of displaying his fall in a less embarrassing way than reality showed it, Knuckles sent another holo journal to his boys – the same ritual he followed every day. Alan pranced around his quarters, admiring himself in full uniform – he flexed a few times, let the disappointment of going home solo pass, then chose another enemy starship to study in detail. Hensaro spoke to his family on Vahoona then prepared what he would say to ease the humans through the experience of warp travel. Dilania prepared herself physically and mentally for what lay ahead, running various fighting simulations through the UT – challenging herself by using different techniques for each attack.

Rhiannon spoke to Walter and Anne, stalling for Cooper's arrival while he talked to Bushaii.

*

"You are right to be concerned Cooper. The Big Two have control of the battle, but things will turn," said Bushaii.

"You do know we're not ready for this."

"That is why we are training away from prying eyes."

Cooper had a thousand questions burning; they stumbled over each other to get out.

Bushaii sensed his frustration, knowing reassurance was required. "We value you and your people too much to send them to battle before they are ready. We have much planned."

Cooper laughed. "I know, I think it will do me good to get there."

"Your presence will do us all good."

"Thanks Bushaii, I best go. I have a ship full of fighters to bring your way."

"We await your arrival, good friend."

*

"Tell us about this place," said Sam Parker from next to the camera lens.

The Ginge breathed in deep and contented. "I love it here... it's like you need a place to go to make the all the craziness of this trip seem worthwhile."

Sam leaned in off camera and smiled to his subject. "And that place for you is the bar?"

The Ginge beamed. "Yeah."

"I see many of the crew using the entertainment decks in their downtime what is it you—"

"Not the entertainment decks, this place – Cocktails & Dreams. This is the place all discerning members of the Dorsano want to be seen."

"Discerning?"

"Single."

"So, it's a place to meet up with other single crew members?"

The Ginge chose a broad smile as his answer.

"So you've had... relations with any of the crew?"

The Ginge shifted in his chair and cleared his throat. "That's a little personal, isn't it?"

"Perhaps, but I'm here to capture the story of the UFO4's time aboard this ship and, since you have a reputation for your success with, as you put it, 'the ladies', it seems like an appropriate question."

"Sure... it's just; I'm not really the kiss and tell type."

"Aren't you writing a book on your galactic... sexual exploits?"

"That's..." said The Ginge, but failed to find any other words to connect with it.

"So, have you added some new material for the book while you've been here?"

"Look, it's not all about notches on my belt. Look around you, this is Cocktails & Dreams, it's about a touch of Earth up here in space, it's about bonding – connecting with people, it's about the atmosphere, it's about the vibe. Don't cheapen it... you're better than that."

Sam smiled to himself, then consulted a clipboard before continuing. "So, before I move on, what do you say to reports that you've... and I quote, 'tried it on with every female aboard the ship'?"

The Ginge went red. "Who said that?"

Sam didn't respond, leaving The Ginge's words hanging in the air.

The silent judgment was too much for The Ginge to bear. "It's a tough crowd. Most of these chicks are, like, five, ten years older than me at least – they're smart, beautiful, talented – well out of my league. If I did pull, I'd be in the, like, legend category."

Again Sam eyed his clipboard and smiled.

The Ginge looked around at the crew, sensing he was the butt of a future editing joke. "Have I showed you the virtual deck? I love hanging out there too. You know the 5S games I design, right? What about the training decks? Sports levels?"

"That sounds interesting," said Sam. "Maybe some other time."

"But it shows my intellectual side, I—"

Sam consulted his clipboard, flipping the page. "We're booked out all day. You're friend Pete is next."

The Ginge went to speak but words failed him again. He sat at the bar as the crew packed up around him. He couldn't escape the feeling that he had been used but he couldn't put his finger on how or why. He thought about getting in touch with Pete to warn him to keep his guard up around the documentary team but then thought if they were setting the boys up it might be better for Pete not to know, so he wasn't the only one looking stupid. Instead he turned to the bar and ordered some hair of the dog.

***

CHAPTER 17

Commander Hank Reynolds sat in his quarters, waiting for the arranged call to activate his communicator. At the moment the clock ticked over to 10.30am GMT the interface bounced into life – it was Oil Man – and on instruction his hologram zwipped into existence on the floor in front of Hank.

"Greetings, sir," said the commander.

"Anything to report?"

Hank didn't flinch at the lack of pleasantries; he was never one for small talk anyway. "We are underprepared, sir. As expected. It looks like the flight simulations are on hold until we reach rendezvous. We will be combining resources with the other life forms and undergoing a unified training regime."

"How is the leadership?"

"There is a lot to factor in, we are—"

"The no-bullshit version, Commander."

"Sir?"

"What is going on?"

Reynolds exhaled heavily through his nose. "I think they're in over their heads, sir. The training feels underprepared, they've got crew dressed up like… well, it seems more eye-popping than practical and last night the captain and his cronies drank way too much with the rank and file. They appeared to be more interested in showing of their skills with the UT and getting laid than setting the right example, sir."

Oil Man pondered the information he had been presented with, nodding absently. "Thank you, Commander. And your team? In position?"

"We have three in the command crew. The rest scattered throughout the ship. Eyes and ears everywhere, sir."

"Excellent. Good work, Commander, keep me chformed."

"Yes, sir," he said as he saluted.

Oil Man reciprocated before his hologram disappeared from view.

*

Sam Parker nodded to Pete that he was about to begin the interview. "So, big night last night, how are you feeling?"

"I've been better," said Pete.

"Do you want to talk us through the accident?"

"I wouldn't call it an accident as such."

"Maybe it would help if you could see the footage again?"

Pete desperately wanted to say no, but knew it was good content for the documentary and the UFO4 brand in general, despite being anything but a value-add for the Pete brand. He'd already seen a UT replay of the moment he hurt himself – frame by dignity-deleting frame. He sighed and nodded his approval.

Sam replayed the vision and made a running commentary as they filmed Pete's reaction. "After several hours of partying it appears you are deep in conversation with some of the female members of the crew—"

"Can we skip ahead?"

"Does that make you uncomfortable?"

Pete watched the image of himself swagger around in front of the ladies. He cringed with each step, each over-enthusiastic laugh, each way too thought out pose. "Just... fast forward."

Sam complied, moving the footage along to the moments before Pete took flight. "It appears you and senior tactician, The Ginge, are exchanging words in front of the crew members."

"Typical Ginge, trying to show off for the ladies."

"Then you produce a small cylindrical object from your pocket. What exactly is that?"

"That's a bullet – Ginge claims it was given to him by Tom Cruise."

"Whatever the case, he seems to desperately want to get it back. He throws himself at you. Do you want to talk us through what happens next?"

"I thought I'd jump skyward and hover to where he couldn't get me."

"That's where the vision shows you flying."

"Kind of. It's a trick with the UT."

"How exactly do you perform this trick?"

"I surround myself with a thin film of UT matter, then will myself to move upwards and stay there and the UT obeys."

"At this point the entire crew are watching and cheering. You appear to be showing the effects of alcohol consumption as you excite the crowd into a frenzy – moving through the air with ease. Then you seem to misjudge a sharp turn and, well, watch for yourself."

Pete looked at the hologramatic replay, while trying to not let too much emotion show for the cameras on him. He watched as a drunken version of himself hovered above the bar, using hand gestures to incite the crowd into larger and larger levels of delirium at his apparent superhuman skills.

Hologramatic Pete pointed out one of the girls in the crowd and grinned like a buffoon. In the next instant it all fell apart. He made a sudden turn; found himself heading for one of the lights hanging down from the ceiling, adjusted direction again only to spiral down towards the crowd. That's when he panicked. All he had to do was tell himself to stop and he would've hovered on the spot, all momentum gone. Instead, he tried to roll around and fly out the other end of the manoeuvre in triumphant control.

But holo Pete pulled out of the move at the wrong moment, turned again sharply to avoid another member of the crowd and was greeted with a face full of front bar. The collision was as spectacular as it was loud and painful. The crowd went from a collective cheer to a collective scream followed by a hush of uncertainty. Then Pete's feet could be seen starting to move from within the hole in the bar he'd created, and one of the bartenders confirmed his survival with a triumphant, 'he's alive'. The crowd moved from uncertainty to laughter and Pete, watching the scene unfold again, felt his face go burning red.

He thought about how best to respond, eventually settling for the humbling, "not my finest moment."

Sam sat and watched, giving a look of disdain. It didn't do anything for Pete's camera confidence. When he'd seen Pete squirm enough, he hit him again.

"Do you have anything to say to The Ginge? After triggering your accident, I figured you'd like a chance to comment—"

"What are you talking about?"

"If you'll permit me to play the accident again from another angle I think you realise what I mean."

Pete looked at Sam, then off-camera to Jason. "Really?"

Jason nodded. Pete still felt his face burn but it was no longer through embarrassment but through the lack of control of what was happening with the interview. He was the one who should be behind the camera, calling the shots or telling the story.

The holo-replay started again, this time the perspective was from over The Ginge's shoulder. Pete was hovering in the air, stirring the crowd into a frenzy while The Ginge was yelling insults. His words were drowned out by the crowd eating out of Pete's hand. The Ginge swore, then turned to the table of drinks, reached into a jug of drink, removed a chunk of ice and threw it at Pete.

The ice missed but not by much. Close enough, in fact, to cause Pete to instinctively change direction. It was the move that triggered the chain of events the left Pete flying face first through the bar.

Pete searched his mind for memories of ice flying by – but he drew a blank. He felt himself going red a third time – this time through anger. At The Ginge.

"How does that make you feel?" said Sam after another sustained pause.

Pete wanted to strangle The Ginge but he knew he had to present a front to the camera.

Sam watched the conflict unfold through the expressions on Pete's face. "It's not really the behaviour you'd expect from a senior crew member aboard a starship... or from a friend."

Pete felt the walls of control close in on him. He was being set-up, he was being baited for a reaction by the very documentary team he helped hire, that he was paying. Worse still, his manager, Jason, watched on as if in on the move. He was trapped and alone in front of the camera.

He glared at Sam, turned and offered the same look to Jason. Then he reached under his shirt and ripped of the lapel mic, stood up and left the room.

Jason nodded his approval to Sam then followed Pete to smooth out the situation.

*

Rhiannon sat with Anne Martin and Walter Johnson in the situation room asking them about their flight. The table was adorned with an abundance of refreshments but they both opted for water – dehydrated from the shuttle trip. Cooper materialised in the room, apologised for the delay and took a seat next to Rhiannon at the large circular table. As grandiose and practical as the table was for big meetings with large numbers of people, it seemed awkwardly ill-equipped to cope with smaller numbers – sitting near each other didn't offer the best view and sitting opposite seemed ridiculous.

Cooper joined in on the small talk for a minute and the rapport between Anne and Walter was immediately apparent to Cooper and Rhiannon. They had only met less than two hours ago but whether it was the unusual circumstances, the exotic transport or the intriguing secrecy – they'd clearly bonded. Maybe they just clicked, as people do; either way, it was a win for Cooper.

When he felt the conversation had taken its course he changed angles. "As you know, we're departing shortly."

"I bet you're excited?" asked Walter.

"…and terrified."

"You'll be fine. What you've built here is amazing."

"Which is only the start of what we plan to build. In fact, that is why you are both here."

Walter and Anne exchanged glances while Rhiannon handed them dossiers and Cooper continued talking. "After we leave, it will be announced that we are about to roll out a fleet of starships, based on the Dorsano's design. A team will be put together to build a permanent moon base – it will be a factory for starships large and small.

"The concept was part of the negotiations with the world's leaders. It's the ideal project to prepare our world for the Universal Translator. With the UT and creators, the ships will essentially build themselves but the collaboration on design and fit-out will be a great step forward for us as a species. It will fast track Earth's understanding of the UT and help people, businesses and governments prepare for the future.

"The goal is to have the moon base complete, structures in place and a team ready to go to be in production of craft within six months—"

"Well, that's impossible," said Anne.

"Which is why I want you to lead the project. You helped put the Dorsano project together, you can help here too."

Anne eyed Cooper. "Have you any idea how lucky we got with the Dorsano? The only reason this project got up is because you had such a massive upper hand in negotiations, wha—"

"What do you mean the only reason?"

"Do you really think they wanted you running things? And with you out of the picture every country and corporation in the world will want to get a piece of the moon base action, no doubt. It's going to be a corporate and political free-for-all."

"Which is why I need you running it. Look, Anne, right now there are only a handful of people I trust beyond my family and the boys. You understand the game, you proved yourself on the Dorsano and I can't think of anyone else I'd want running the biggest project in human history. "

Anne looked at Cooper, then Rhiannon, she flicked through the dossier and contemplated the offer. "It sounds like a disaster waiting to happen to me."

Cooper knew Anne well enough to know her tone was interested, he watched her facial expression change several times as she pondered her future.

"Very well, but we run it my way."

"Brilliant!" said Cooper. "And absolutely, yes."

Walter's eyes darted about as he poured over the information Cooper and Rhiannon had provided. "So, how do I fit in to this picture?"

"That's another project altogether. I want you to begin work on another moon base, underground on the far side of the moon."

"What? Why?"

"What Anne was saying is true. I don't trust any of them. Not with the fleet or Earth's best interests, and certainly not with mine. This base will be secret, known only to us. You won't be restricted by governments or corporations and you'll have the full capabilities of the UT at your disposal. You'll be able to build quicker, far quicker."

"I don't know how to use the UT."

"Before we leave I am issuing you and Anne with full interface rights, you'll have complete access to it – the same level as me. No one teaches you how to use the UT better than the UT itself.

"Look," added Cooper. "Should everything turn out to be above board and this is just paranoia on my part, well, we'll have extra ships ready to go. But if something were to happen to us up there, or with the powers that be down here, we've got insurance."

Walter nodded knowingly. "I like your thinking your Mr Simpson, I'm in."

"Sweet," said Cooper, instinctively offering a high five to Rhiannon, then awkwardly to Anne and Walter in turn. "Gandalf, are you there?"

"I am everywhere."

"Can you please send me two communicators?"

The devices appeared on the table in front of Anne and Walter.

"Fit them around your wrists," said Cooper. "You can contact me anytime."

Anne and Walter complied then played with their new toys.

"Gandalf," said Cooper. "As discussed I hereby grant Anne and Walter full UT access."

"It is done."

"Thanks Gandalf," he said before turning to his new recruits. "Thank you both. Before I go, do you have any questions?"

"More than the time you have, I fear," said Anne.

"Get yourselves home, establish your connection with the UT, send me your questions and start your plans. Work together but stay distant. We can't tip off anyone else as to what we're doing."

*

Knuckles sat in the observation deck, facing the wall to ceiling view of the Earth slowly turning in front of him.

"Look, why are we doing this?"

Jason sensed his apprehension and aimed for a consoling tone. "For the documentary, Pete told me you'd been briefed."

"He told me you guys wanted to interview me but I didn't think it would be like this," said Knuckles as the crew busied themselves setting up lighting and camera equipment.

"Are you thinking about the messages from home?"

"Yeah, it's private. That's why I come up here."

"I understand. What Sam and his team are trying to do is show the real Knuckles."

"What?"

"Well, the world knows your character from the movie, the guys are trying to show something... more intimate. What makes you tick, that sort of thing."

"Yeah, that's what Pete said, the thing I don't get is why you want me watching the video diary from the kids."

"That's exactly the sort of thing that helps people relate to you."

"And it's private!"

Jason headed to the window and looked out over the view. "Did you know you were the least popular of the team?"

"What are you talking about?"

"We've done testing, a lot of testing actually. It helps us focus what pillars of your branding we—"

"You lost me at branding... or was it pillars... or testing. I don't care about any of that."

"Sure, marketing speak aside, what fans are saying is they like you, they just don't feel they know you as well as the others." Jason held his view over the window. "Seeing you in a personal moment with really help with that, it's going to make great viewing."

"I'm sure that all works well for you and Pete and what's his face over there, but—"

"Sam," said Sam.

Knuckles paced left then right. "It's easy for you. I'm just ratings for you. I'm just dollar signs. Twelve percent."

"Sixteen percent," correct Jason before he could stop himself.

Knuckles stared at him in disbelief before he continued his caged lion pacing. "You've got no idea! You know, I was this close to not coming back to Earth. It's funny, when I was gone and it all got too much, the only thing that kept me going was getting back home. But when that chance came I realised it wasn't ever going to be what it was.

"Do you have any idea what it's like to be the laughing stock of an entire planet?"

"You're not a laughing stock Knuckles, you're a star."

Knuckles stopped pacing and laughed mockingly. "Thanks... but I'm pretty sure the sixteen percent might have something to do with that opinion. I know what people think. I know what they say."

"I've never met an A-lister who didn't have haters."

"I'm not an A-lister... look, you're missing the point. I come up here for the peace and quiet. Every day I get a message from the boys, every day I send one back. I can't be with them, I ... I don't belong. I don't belong on Earth either."

Knuckles made a noise, neither a laugh or a cry. "Maybe this is the only place I do belong... alone."

Jason, studied him. "I get that. Look, Knuckles, you're my client... but you're more than that. And these guys, they're here to tell your story, not to embarrass you. I know what this interview would do for telling your story, for helping the people from your home world understand."

Knuckles walked to the viewshield and gazed out over his home world.

"Look, we don't have to do it all today; we can work up to it. What about, for today, you let Sam and his team film how you communicate with the kids?"

Jason saw enough in Knuckles expression to know he was contemplating the idea. He turned to Sam and gave him a nod – they knew they could already turn Knuckles words into an interview. "Sam, can you move the cameras back and film Knuckles fly on the wall style? Nothing intrusive, no questions; just observation."

"Can do, something discreet could work really well here actually."

Jason turned back to Knuckles, seeing his mind tick over. "You just have to do your thing and forget we're even here. We'll take care of the rest."

Knuckles maintained his focus on the Earth.

Jason moved in next to him and cast his eyes over the hypnotic view. "Tell them your story. Not for us, for you."

There was an age of silence before Knuckles nodded. "Alright."

"Yes," said Jason before moving in to hug Knuckles. He soon realised the move wasn't being reciprocated and switched it to a high five. It too was denied so he held out his hand for an old-fashioned shake and when that wasn't met he awkwardly patted Knuckles on the back. "Well done."

It had been less than a minute and Knuckles regretted his decision.

*

Cooper looked down the barrel of the camera, his eyes squinting in the bright light. "It just feels a little weird, like, not natural."

"Can we get something to tone down the light?" said Jason.

Sam Parker scratched his beard as he studied the scene. More accurately Jason, who continued to be more hindrance than help. But he was about to interview the captain – the money interview as he called it. He knew he'd need to weed the middle man out of the communication loop to get eh most of his subject matter. He walked over to where Cooper sat and looked back at the camera. He shielded his eyes in sympathy. "You're not wrong, Cooper." He gave his interviewee a friendly nudge and Cooper reciprocated with a thin smile.

He moved back in front of his subject, using his fingers to frame him in shot, before retreating to the camera to make adjustments. Then he moved to his seat next to the tripod and grabbed his clipboard. "Your eyes will adjust soon enough, Cooper. Try to forget about the cameras, we're just two people have a chat."

Cooper looked at Jason then back at Sam before nodding his agreement.

Sam saw he still had some work to do to cut the cord. "Water?"

Cooper nodded and Sam signalled one of his assistants who awkwardly stepped up to Cooper with the refreshment. The captain opened the seal and took a large swig.

"If it makes you feel any better, I've had movie stars and presidents sitting in that chair who've looked far less comfortable than you."

Cooper smiled.

"Maybe they knew I was coming after them," said Sam before laughing to himself.

A few of his crew members laughed as well. Sam set his sights on Cooper again. "Don't worry, this is a friendly chat."

"Although rumour has it you have some good oil on the American President. Might have to have a little chat with you about that one day."

Cooper raised his eyebrow suggestively and they shared a smile.

"You've done well, you know," said Sam.

"With what?"

"This ship – everything. It's... amazing."

As Cooper responded Sam smiled, in part to engage, but mostly because had his interviewee where he wanted him. They exchanged a nod that meant the cameras were rolling. "So, how does it feel to be sitting aboard your spaceship, Captain Simpson?"

"It's... I'm not sure... surreal. It still feels like a dream. Ask me again in a few weeks, maybe."

"In a few weeks this ship may well be at war."

"Yeah, that's... that's a pretty cold fact."

"That must be something you find difficult to imagine."

Cooper laughed to himself. "I don't think I can imagine what that will be like. How could anyone? Everyone on the Dorsano is new to this. I guess we'll be learning on the job."

"What are you hopes for this mission?"

Cooper paused. "This is the start of a new era for humanity. I want to show the galaxy how helpful we can be. I want to help us win then return home to help Earth and our people transition to become a species of the galaxy."

"You certainly have done the unimaginable in putting the crew together in the timeframe that you did."

Cooper studied Sam, sensing a shift in tone. "We had a lot of help."

"Do you feel you've got the right people?"

"The crew? Yeah, absolutely. We've hand-picked the best of the best from around the world."

"So, you've hand-picked 1500 people in a few days?"

"I... well, no, not me personally but—"

"So, how can you even vouch for those you'll be fighting with?"

"We've been in constant contact with the world's governments and military, we have the most talented fighting force Earth has ever seen."

Sam nodded his approval to let Cooper know he'd passed the first test. "Tell me about the gamers, you have 35 people on board whose only apparent skill appears to be playing games."

"Absolutely, some of the most interesting crew members too."

"What does someone who's good at Donkey Kong bring to the mission that, say, a seasoned fighter pilot, experienced in using state-of-the-art hardware, can't?"

"For a start no one is on board for being good at Donkey Kong. Perhaps if you'd picked up a game in the last thirty years you'd know they are very advanced. Shooters, real-time strategy games, flight sim, RPGs – and the guys playing these games at the highest level are amazing."

"And they'll help you win a war."

"You've got to understand what it's like up there, Sam. I know everyone has seen the movie and knows about the Universal Translator. But it's one thing to see it; it's another to experience it. It's not an app you interface with, like something on your phone, it becomes a part of you – you interface with it constantly. It's deeper than that really, it... intertwines with your sense of reality – the two aren't separate – they're one. There's something about a gamer's mind – they get it. And on this mission, we need people who get that as much as we need people who understand command structure or people who are trained killers."

Finally Cooper had a few seconds breathing space as Sam recollected his thoughts.

The interviewer gave Cooper another nod before reloading. "So, tell me about the uniforms."

"That's probably a question for The Ginge."

"But you're the captain, what's your opinion?"

"I was a little like Rhiannon—"

"Rhiannon Richardson? Your partner?"

"Yeah. She was apprehensive at first but she's gotten used to it."

"Do you want your team to wear something they have to get used to?"

"Most haven't had to – they love the design. They look great on."

"You don't find them a little... revealing?"

"Perhaps? If I was looking through conservative, early 21st century human eyes, I might. But we are an elite fighting force, we are going to join a galactic community that are far more liberated and assured than we are. That is where we are heading and if I'm looking through my 'Earth as part of the galaxy' eyes, well, the uniforms are completely normal."

"Normal?"

"It's like the designers said, 'we're breaking paradigms'."

"Well, no one can argue that your paradigm-breaking crew don't look the part, but can they act it? How is the training going?"

"I'd call it a work in progress. We're... we're coming from a long way back. Sure, we've got good people; talented people; but flying a fighter jet and getting behind the controls of an attacker are completely different things, you're dealing with zero gravity, new flight systems, controls, weapons – that's a lot to learn in a short space of time. And that's not to mention the communications and other systems everyone has to adjust to... it will come, it will just take a bit of time."

"But it seems to me that time is the one thing you are lacking."

"True, we leave Earth shortly, but we are going to hone our skills with other, more experienced, species. We'll be a well-oiled machine by the time we see action."

"There are some people who say you're not up to a situation like this, what do you say to them?"

"Who would be right for this? I know I've been representing Earth before any of them even knew there was intelligent life beyond our world. And I think I've got some runs on the board that no one else has. Am I perfect? No. Am I a bit young? Maybe. But I've got the experience. I care about the Earth and its people and I want to be a part of the new Earth. This situation may have chosen me first, but I chose it now."

Sam looked at his subject and nodded. "That will do for now, Cooper. Well played, son."

Cooper felt a wave of relief wash over him. He shook Sam's hand, nodded to Jason then set his mind back to preparations.

*

"Coops, Coops! Just the man," said Pete as he jogged up to the captain, skidded to a halt with beads of sweat glistening on his forehead.

Cooper eased his quick pace to a slow stroll, refusing to give up all his momentum if the conversation didn't warrant it. The interview had slowed him down enough for one day. "Not a good time, Pete."

"There's never a good time with you these days."

"It's called being a captain."

Pete looked him up and down. "It's called being a douche."

Cooper picked up to hustle speed again.

"Coops!" said Pete again, in a tone that was the perfect combination of squeal and panic, ensuring the attention from the crew members in the immediate vicinity was drawn.

Cooper stopped, observed the minor scene they were creating then glared at him. "Seriously? Just... make it quick."

"Sure thing," said Pete before settling in to epic movie pitch mode.

Cooper rolled his eyes as a pre-emptive strike.

"Picture this. Deep space fills the screen, dotted with a wash of vivid stars and a colourful spiral of gas cloud. Underneath the galactic eye-candy and subtle stirring of an orchestral score simmers away, hints of the main theme play, but just hints, teasing future moments of wonder and majesty. The camera pans—"

"I said quick."

"Relax, I'm just trying to set the scene."

"For what?"

"The idea I had."

Cooper started into a stroll again. "OK, sure, I'll humour you. What idea?"

"So, the history of adventure and discovery in books and on screen are filled with leaders taking their ships and people into the unknown. People of vision, people of—"

"Can we get to that bit where you tell me your idea then leave me to do the captaincy thing?"

"That's exactly what I'm talking about."

"I'm lost."

"The captaincy thing – a captain needs to engage with the viewer on a personal level – his thoughts and plans. So they can get a mainline into the big picture of the mission and see the man behind the badge."

"I have no idea what you're talking about."

Pete breathed heavily in frustration, collected himself then continued. "Deep space fills the screen, dotted with a wash—"

They reached the lift that would take Cooper down to the training facilities by the docking bay. Cooper interfaced with the holodisplay, summoned the lift, then realised there would be an agonising few seconds before it would arrive – he was stuck with Pete. "Make your point!"

"OK, so, fast forward the scene setting to where a voice echoes out over the gentle score; oboes, I'm feeling oboes here for some reason. And violins, obviously, maybe a lone female voice—"

"Pete!" snapped Cooper as the lift arrived.

He got in and Pete followed. Cooper avoided eye content, knowing giving it would only encourage the talky man further. "Well?"

Pete put on the most majestic voice he could muster, which wasn't very majesticy. "Captain's log, stardate... whatever the stardate is... make it up for all I care. Our mission, to join the frontline of the New Council defences, to face the enemy in battle—"

"What are you doing?"

"...Morale remains high, but today I know we'll be tested like never before. I'll be tested like never before. I pray we have learned the lessons that will lead us to victory."

Pete let the last word hang in the air and stared at Cooper. It was too much for the captain, who eventually had to return the look. When he did he saw Pete beaming at him, a knowing glint in his eye.

"What was that?"

"A captain's log, well, your captain's log. Well, not necessarily your captain's log specifically, you'll obviously bring your own thing to it. But something like that."

The lift opened and Cooper flew out, Pete in tow. "We have been talking for three minutes and I still have no idea what you're saying to me."

"I want you to do a captain's log, of the mission – a document of your thoughts, ideas and things of note," said Pete, struggling to keep up.

"No way!"

"I knew you'd say that. That's why I did the scene setting stuff. Think about it, it'd be the perfect device for the viewers."

"Except, I'm not doing it."

"You have to... all the great captains through history have."

"Like who?"

"Captain Kirk, Captain Piccard..."

"Apart from on Star Trek?"

"Heaps of other captains."

"Name one."

"Umm... erm... Captain Pike!"

"Who's he?"

"Just some sci-fi movie you probably haven't heard of."

Cooper could see his goal in sight. The training facility and safety beckoned. "Is it Star Trek?"

Pete swore under his breath. "A little bit... maybe."

Cooper reached the training room door and stopped and looked at Pete. "Look, I'm not doing it. It's lame for a start, secondly, I just did an interview and, thirdly... I'm just not doing it."

He turned to leave but Pete grabbed his arm. "OK, forget about the captain's log thing. How about this, just start a video diary. Just a personal record of what you're going through while running the ship and going to war. You're creating history here Coops, record it."

"It sounds a bit like the captains log thing."

"All I'm saying is do it for yourself. When we get back to Earth, have a look through what you've posted. If you're happy to share it, I'd love to use it in the movie, if not, keep it to yourself. Simple."

Cooper looked Pete up and down as he mulled the idea over. "I'll think about it. Now can I go to my meeting in peace?"

"Brilliant. Sure thing. Thanks, Coops!" said Pete as he turned to head back to the editing suite.

Cooper headed into the training facility as he heard Pete's voice echoing off into the distance. "Editor's log, stardate... something or other... after speaking to the captain..."

*

Time did what time does best – it went fast, while it took forever. The duplicitous illusionist. The finish line, departure time, barrelled towards Cooper like an omnipresent wall. Every time he looked at it, it would slow to a stop – forever away. But when he buried himself into the work at hand, the wall would sneak closer than ever before.

It was like a watching a magician at work, drawing attention to the right hand, playing you for a fool with the left. Bait and switch.

Time; the duplicitous illusionist.

Before he knew what had hit him, Cooper was sitting in the Cocktails and Dreams bar once more. It was the last night before departure – where had the time gone? So many things had gone unfinished, but he had come to terms with that. Maybe he could've spent his last night tweaking this and adjusting that, but what would be the point? The ship and the crew were simply in the state they were in and a few more hours by one man would not make the difference – not against the external forces that would surely come to bear on them soon.

Around him, the boys, Rhiannon, Alan, Dilania, Hensaro, Jason, Reynolds and crew celebrated and congratulated each other. The documentary crew and the bar staff were the only people working but the sense of euphoria even swept through them. Everyone was caught up in the moment.

Except Cooper.

He'd allowed his body to be present with the others but his mind still pored over the details he'd missed, or what he could do when they reached Baltwae to improve this or that.

He studied Jason and the little look he'd only just started noticing but couldn't let go of – the one that made sincerity hard to believe. Then Reynolds, Oil Man's main man; all military cold and undisclosed judgement. Beyond them, the documentary crew buzzed about the crowd, Sam and his team telling their own story.

It seemed everywhere he looked was someone with their own agenda. Somehow they had all been pressed together on this crazy, unlikely. Cooper could feel the bond of it all, and the weakness in the glue. This wasn't four boys against the galaxy; this was a much more complex beast – a mountain of moving parts, spinning a semi-controlled course into the unknown. He wondered what would happen should it all go awry. Would the parts still hold together?

He lifted his drink to his lips and thought about difference, leadership, pain and victory. He prayed his captaincy could keep the machine together if the worst did happen, then took a swig.

He heard an uncomfortable grunt to his right and turned to see one of Dilania's skeleton crew sit next to him. The impressively large blue skinned being, the HUD reminded him was Gluff Hurn. The chair groaned with displeasure as Gluff found a comfortable sitting position. Then he turned to Cooper and nodded his servitude and respect.

Cooper smiled and nodded back. "Enjoying yourself?"

"Your alcohol makes my legs misbehave."

Cooper laughed. "It will do that."

"But it also gives me the courage to talk to my captain," said Gluff. "And to thank him."

Cooper smiled. The words couldn't have hit him at a better time. For every Reynolds, there was a Gluff, maybe a dozen, or a hundred. From nowhere a tear welled in his eye, he wiped it away before anyone saw. "Your captain thanks you. I'm sure after a good rest here you'll be ready to socialise again."

"Gluff finds he learns more from watching and listening than talking."

Cooper took another swig. "I hear you."

The pair watched the crowd in silence for a while longer before Hensaro approached. "Gluff, I hope you're not bothering the captain."

"He's fine," said Cooper. "He just needed a rest."

"And you, Captain?"

Cooper smiled. "I think I needed a rest too."

Hensaro put his hand on a chair, then looked at Cooper, seeking permission to sit. Cooper nodded. "Of course."

He took his seat and joined in the silent observation both of his captain and the happenings around them. "You make a good captain," he offered eventually.

Cooper smiled uncertainly. "I guess we'll find out soon enough."

"No, I believe we already know."

It was as if Hensaro had read his thoughts and doused his fears in a moment. Cooper nodded his appreciation.

"My apologies if Gluff has made you feel uncomfortable, he doesn't really say much."

Gluff turned to watch Cooper's response.

"No need, he's thanked me and been perfect company since."

"He spoke? A rare compliment."

Cooper turned to Gluff and nodded before the three of them raised their drinks together and took in a mouthful.

"Before we go any further I have a mutual friend who wishes to say his congratulations."

While Cooper looked on quizzically, Hensaro closed his eyes in secret dialog with the UT; soon the image of Bushaii Bartton appeared in front of them. "Cooper Simpson and Hensaro Althot side by side, now there's a vision."

Hensaro offered his commander a hand gesture to acknowledge rank before laughing. "Good friend, I await our reunion."

"We shall celebrate long and hard. I believe I owe my human friend here a large piece of Fredahnian hospitality."

Cooper laughed. "It will be nice to finally meet you face-to-face."

"It is an injustice that it hasn't happened sooner."

Hensaro put his hand on Cooper's shoulder. "I believe we have caught our human friend in a moment of doubt over his leadership skills."

"Is that so? Well, young leader, you could not be further from the truth. In fact, you are a significant part of the fleet we are putting together. A fleet I am confident will be amongst the powerful there is. The Fredahn at your side is one of the finest men I know and he speaks of you in the highest regard."

Cooper looked at Hensaro, then Bushaii's hologram. "Thanks."

"What you feel is nothing experience cannot mend," said Hensaro.

"And spending time with other leaders," added Bushaii.

Hensaro took another swig. "It seems the arrival feast cannot come quick enough for any of us."

"Or war," said Bushaii. "On that note, I must prepare for the arrival of the accomplished Hensaro Althot and the great Cooper Simpson. Until we soon meet."

"Commander," agreed Hensaro.

"Yes sir," added Cooper, before saluting.

Bushaii Bartton's likeness disappeared leaving Cooper, Hensaro and Gluff to their drinks and observations.

"Care for some female company?"

Cooper felt Rhiannon's familiar touch on his shoulder. "Here she is."

Gluff moved his chair across and lifted another into place next to the captain. As Rhiannon sat, The Ginge, then Pete and Knuckles took their seat at the table. Dilania soon joined them, closely followed by Alan.

They talked, laughed, joked and drank. No mention of the morning, no mention of the state of the ship or the crew – just banter and friendship. For a brief moment in time, Cooper became lost in the moment, free from the duplicitous illusionist. He could just be.

***

CHAPTER 18

Cooper looked at his reflection and played with his hair until Rhi told him to stop stalling. It was only then he realised he was doing exactly that. Sure, he still felt like death and lichen, but beyond that, he knew this walk from the captain's quarters to the bridge would be one of the most significant walks of his life.

He could hear a crowd gathered beyond the door to his quarters. He turned back to Rhiannon, "You ready?"

"I'll be right behind you. But I think they're waiting for their captain."

Cooper gave her an are-you-sure look; she nodded. He gathered himself, then faced the door once more, this time making his exit. On the other side the entire crew were lining the route to the bridge. Everyone was kitted out in full Dorsano crew attire – they looked magnificent. The navy blue of the military crew dominated, the sheer volume of uniforms in the non-dominant colour acted like a magnifying glass, highlighting pockets of red, gold, green and azure.

The Ginge, Knuckles and Pete were waiting for him in the middle of the path. The Ginge jogged up to his side beaming his awkward smile to the gathered crowd. Cooper leaned in to shake his hand. The Ginge lifted his hand as if to meet it, only to bypass the physical engagement and continue to his forehead where he saluted his captain.

Cooper smiled awkwardly before reciprocating. "At ease."

The Ginge leaned in close and whispered. "How good is this?"

Cooper looked around; there were no words to describe what he felt, at least none he could grasp.

"Seriously, we look the biz, Coops. I'm totally ready to kick ass."

"We definitely look ready," said Cooper.

Soon Pete and Knuckles joined them and Cooper began the inaugural walk to the bridge, after he'd taken a couple of steps the others followed in his wake.

Someone started clapping, and was soon joined by another. Soon, the entire crew stood in ovation for the historic moment.

"Captain on deck," yelled Jasmine Jennings as they neared the bridge.

Pete, Knuckles and The Ginge tried to hide their laughter under their breath. Cooper looked at The Ginge seeking explanation.

"Deck! Gets me every time."

"I'm lost."

"You know, the New Zealand accent thing. Her E's sound like I's."

Cooper maintained a confused stare at his redheaded friend.

"So, captain on deck sounds like captain on di—"

Behind him he heard Pete and Knuckles snigger again.

"Really? We're about to launch into space for the first time, do you really think it's time for… deck jokes?"

"It's always the time for deck jokes," said The Ginge. He winked at his captain, who rolled his eyes. The Ginge missed the response, already having turned to give Jasmine the same treatment.

Cooper saw the exchange. "Are you and her..?" he whispered as he absently lifted his hand to acknowledge the crew.

"What happens at Cocktails and Dreams, stays at Cocktails and Dreams," said The Ginge, hoping his bluffing was better than his pulling.

Cooper called the other boys into a huddle and whispered, "Game time boys. Seriously, can we, just for today at the very least, try to act like we know what we're doing?"

They nodded then Cooper turned once again to head into his future.

*

The circular space of the bridge opened up before him, revealing the on-duty flight crew at their posts – all but The Ginge, Pete, Knuckles and Rhiannon, who soon took their positions. For the first time, Cooper experienced the bridge fully manned, instrument panels at the ready and striking azure uniforms making the scene the stuff of dreams. Everyone faced him, awaiting instructions.

Cooper made his way along the bank of engineering displays at the back of the space. Heads turned as he went, tracking his movements across the deck. He nodded to each of the crew as he passed. He walked around the main control desk towards the tacticians and gunners at the front of the bridge and soon found himself staring out over the Earth and the stars beyond her atmosphere.

He smiled again – it dawned on him that this moment was too good to actually be happening. Surely, he was dreaming after all.

He completed his circumnavigation of the bridge, acknowledging the other gunners and tacticians before he approached the central control deck.

Dilania, who was occupying the captain's chair, stood. "It is my greatest honour to pass the helm and this craft to its new and rightful captain, Cooper Simpson – a true and worthy soul."

Everyone on the bridge applauded as Dilania stepped back from the seat and offered it to Cooper. He looked at her and nodded.

"A champion of men, a hero, a wise soul with a gentle, yet assertive touch—"

"Thank you Dilania," said Cooper before glancing briefly towards Rhiannon.

"Your hospitality, help and expertise are greatly appreciated by everyone here and on Earth."

He found himself standing above the seat; he looked around the room before lowering himself down. Another impromptu applause broke out and Dilania made her way to her new spot in the tactician's pod. Soon the noise abated and all eyes were on him, awaiting orders.

As Cooper opened his mouth to speak, an interface burst into life in front of Chief Tactician Daniela Rodriguez. She looked at Dilania, who nodded back. "Captain, you have an incoming communication... make that two."

"Thank you, Rodriguez. Rumour has it we might be getting a presidential send off. Display."

Dilania looked at Daniela to offer assistance but the latter shook her head, wanting to do things herself. "Erm, they're holograms, calling the first one up."

The likeness of a human was projected into the floor of the bridge in front of the crew. The figure was unusually hunched forward and surprisingly dressed in a Dorsano uniform. More surprisingly being was a familiar one and, most surprisingly of all, he was already in the room.

"Um, Knuckles... erm, I mean, Gunner Thomas," said Cooper.

"Wassup?"

"You're supposed to call him Captain, numbnuts," said The Ginge.

"Ohh, soooorrryyy!" said Knuckles before adding. "What's up, Captain Numbnuts?"

He laughed but he had no allies in hilarity.

"Do you mind telling me why a gunner has opened a direct holocomm to the bridge?"

"What?"

"Turn around."

Knuckles turned to see his likeness projected in the centre of the bridge, echoing his movements live. He looked at his captain, then around the room, the hologram mirrored his movement. Then he turned his attention to the control panel in front of his turret that had betrayed him. "Buttons," was the only word he could muster in his defence.

Cooper felt himself go red as the uncomfortable silence in the bridge was in danger of undermining his captaincy before it started – if he couldn't keep Knuckles in check, what hope for a crew of 1500? "Perhaps leave the buttons to the big kids. Daniela, can you display the other holocomm?"

The hologramatic likeness of the president appeared in front of the control deck. She, on behalf of the US, and other nations, wished Captain and crew Godspeed for the incredible journey ahead.

Cooper felt both in and out of the conversation. He was there, saying all the right things, ticking all the boxes, but part of him was having an out-of-body experience.

It was as if he was floating above the bridge, watching himself be part of the machine – the machine the Oil Man ran.

Maybe he was just thinking too much but, either way, he knew he was one conversation and a few instructions away from putting a few light years between earth politics and the Dorsano, and he was very OK with that thought.

It wasn't long before all ceremony was completed and the flight crew looked at him with anticipation. He studied their faces and hoped he had whatever it was everyone was hoping he had. He looked at Rhiannon, who gave him a reassuring you-can-do-this smile.

"Erm, I guess this is it," he said. "Is everyone ready to take this baby for a spin?"

"Engineering ready," said Gayan Da Silva

"Tactical ready," said Danila Rodriguez

"Military ready," said Hank Reynolds.

"Flight crew ready," said Pete Bates.

"Excellent. Rodriguez, can you open comms to the ship," said Cooper.

"Affirmative. Comms open in three, two, one..."

"Fellow crew members of the Dorsano, I'm not sure what advice to give you to prepare for hyperspace travel – it's hard to describe and, well, can take some getting used to."

He looked at Pete, who avoided eye contact. A few of the crew sniggered but the thought of facing their own bad reaction limited it. "I can tell you to prepare for the bizarre... there's no shame in having a bucket on standby. In fact, we've decided to pull out of hyperspace a little short of our destination to give everyone a chance to recover before we meet our allies – some of you will thank us for that later. But, enough of me gabbing on, you ready for a ride? I'll see you on the other side... and the other side of the galaxy. Captain Simpson out."

Cooper nodded to Rodriquez, who closed comms accordingly.

"Let's do this thing. Flight crew, coordinates set?"

"Set," said Pete.

"Very well...." said Cooper before it dawned on him how important the next word, or words, were. Many a famous sci-fi captain has uttered a unique phrase to activate warp drive – engage, activate, energise – so many options. But he had to settle on something that was right for him. As everyone looked on, he rifled through potential options in his mind before settling on one that seemed to feel right for him.

"Fire it up," said Cooper.

He immediately regretted his choice.

Pete gave him a double take before responding. "Yes, sir."

***

CHAPTER 19

Experience – all the research in the world can prepare you for the theory, but nothing can make you ready for the reality. As the Dorsano ripped through the fabric of space-time, the vista through the bridge's viewshield was replaced with a surreal kaleidoscope of Dr Whoesque colour and wonder. It wasn't just the view outside either, the bridge seemed to turn a colour palette that made 1960's attire seem bland.

For Cooper, The Ginge, Pete, Knuckles and the aliens, the sensation was one part of a series of steps, necessary to get from one point in the galaxy to another, within less than a lifetime. They knew the feeling – the weightless giddiness in their stomachs – would pass. For the rest of the humans on board the Dorsano it was a different matter. While each person would experience variations of colours, the physiological and psychological test for the uninitiated was like a mixture of torture and hallucinogens.

As the journey concluded and a tangible, material reality returned, Cooper looked around the bridge. He noted the difference in reaction could best be summed up by two people – Hank Reynolds and Alan Woodcock. They were both around the same age but that's where the similarities ended. Hank was all sharp lines and military discipline while Alan was soft edges and pasty geek. Yet it was Hank excusing himself from his post and making a hasty exit from the bridge with his hands over his mouth.

Hank, one of Oil Man's plants, had offered little on the bridge beyond judgmental stares and Cooper enjoyed the beautiful irony that hypserspace travel had left the experienced, clean cut former fighter pilot a quivering wreck, while someone who owned the show Cleopatra 2525 on Blu-Ray was seemingly unaffected. Cooper tried not to let his smile show.

"Whoo-hooo," screamed The Ginge at the top of his lungs. "That never gets old."

Several other crew members chased Hank to the exit. Cooper made eye contact with Rhiannon, she nodded to him. He smiled, impressed.

"Congratulations all, you've officially earned your 'I survived hyperspace' T-shirts," said Cooper.

"Erm, Captain," said Luca Hoffman.

"Rodriguez, you good to put me on communic—"

"She's just... collecting herself," said The Ginge.

"I understood she was about to regurgitate," said Hensaro.

"Thanks, Ginge can you put me on ship-wide comms?"

"Coops, we've got a problem," said Pete looking at the data Luca had alerted him to.

The scene through the viewshield was an endless sea of stars. The dark disk of the planet Baltwae could only be made out against the stars it obscured. Even that effect was masked somewhat by the light of the ships gathered to meet there.

Then a pulse of light puffed in and out of existence in the distance, followed by another and another.

"Ahh... what's that?" said Cooper.

"Something's wrong, Coops," said Pete.

"What do you mean, wrong?"

"It's Baltwae, something's not right."

Cooper looked at Pete and Luca by his side, inspecting their displays. "Really? Perhaps you've some detail to expand on what 'wrong' or 'not right' might mean."

"Erm... looks like fighting," said Pete.

"What fighting?"

"Ahh, spaceships fighti—"

An emergency display lit up on the main control panel in front of Cooper and the other senior crew.

"Proximity breach," said Luca.

"What? Where?" said Cooper as he scanned the skies looking for the offending object.

"Not sure. Distance 5km, coming straight at us, impact in 20 seconds."

Cooper became instantly aware of all eyes upon him, some seeking direction, others sitting in judgment over his every action. "Got it. Gandalf, can you visually highlight the threat for us?"

"It's a ship," said Pete after studying the 3D image Gandalf had overlaid across the eerily dark sky. "We'll part of a ship."

"Suggest we test out our firing capability, Sir," said Hank Reynolds, now back on the bridge with some colour returning to his face.

"Ten seconds," said Luca.

"Negative," said Cooper. "I don't want us making our presence known on Baltwae until we find out what's going on."

Hank seethed while Cooper turned his attention back to the battlefield display. "Gandalf, do we have some sort of tractor beam?"

"Yes."

"Five seconds!" said Luca.

"Can you move the object to fly safely past us?"

"It will be close," observed Gandalf.

Cooper and crew braced for impact as they watched the 3D highlighted object move further and further away from the centre of the field of vision. With it came the awareness they had no true sense of how big the Dorsano was – no sense of how much they needed to avoid an object by to not hit it with some part of the craft. Fortunately the bit of spaceship debris proved a good example as it passed safely by the Dorsano's hull.

There was an audible exhale around the bridge. Cooper looked at the crew in turn, making sure he held eye contact with Hank after winning a little battle. Disdain was written all across the military commander's face.

There was a thud on the viewshield.

"Ummm, Cooper?" said Knuckles.

Right in front of his turret the body of a creature, an alien – well, some significant parts of the body of an alien – smeared across the viewshield before sliding across it then disappearing, once more, into the void of space.

"Anyone else hungry?" said The Ginge.

"Ewww," said Pete before adding, "nasty," in a far cooler voice with a mind to his video editing.

"Excuse me, Gandalf," said Cooper. "Any reason that corpse didn't raise a proximity warning on the equipment?"

"The velocity of the object was in no danger of harming the Dorsano."

"Thank you. Can we add a corpse proximity warning as well, otherwise I feel we may be in danger of damaging the bridge's carpet."

"Yes."

"Thank you. Now can we get eyes on what's happening at Baltwae? Rodri—, I mean Ginge?"

"Standby."

The Ginge interfaced with the UT in silence, seconds later the front of the bridge filled with a hologramatic close up of the not-right events near Baltwae. The carcases of ships large and small lined the dark planet's orbit. One large craft, cut clean through the middle, had its back half collide with another ship while the front half slowly picked up momentum and heat as it made its final flight – an uncontrolled descent to Baltwae's cold surface.

For some time those on the bridge remained silent, coming to terms with the level of destruction before them. The entire fleet they had been about to join, their part of the attack plan, their way into the war had been completely decimated.

"Gandalf, what happened here?"

"There was a battle."

"But this place was supposed to be secure."

"It appears not."

Cooper looked at the crew once more, once again all eyes on him, while he thought on his next move. He dialled Bushaii Bartton through his communicator – no response.

"Gandalf, is Bushaii Bartton OK?"

"I am unable to answer that at this time."

"Where are the enemy?"

"They have since departed."

"How long ago?"

"We have missed the confrontation by minutes."

"What about other survivors?"

"There are many."

"Alright, listen up people, we're going in. Pete, get us in as close as you can. Hank, have your teams on standby. I want every craft we have capable of search and rescue in the skies as soon as we get into position. Which will be how long, Pete?"

"A few minutes, depending on how close we can get."

***

PART 3: WAR

CHAPTER 20

"Can you squeeze us past the dark grey one?" said Cooper as he and Pete tried to Tetris the Dorsano into a safe orbit around Baltwae. They were honing in on what Gandalf believed were a number of survivors trapped in the remains of the starship Gaujillo.

"Which dark grey one?" said Pete as wiped the sweat from each hand in turn.

"The giant dark grey starship we're heading towards."

"Yeah, the one on fire or the one that looks like a skeleton wrapped around the tower of Pisa?"

"Ah, didn't see that on-fiery one, I mean skeletony-Pisary."

Pete thought for a second then consulted with Hoffman. "Possibly. It's gonna be tight though."

The proximity alert went off again.

"That better not be another corpse," said Pete.

The Ginge looked at the 3D display. "Could be, it's small enough although the shape doesn't seem familiar. I might use the UT to scope in for a visual."

"Coops, we've got another seven," said Knuckles from aboard one of the rescue craft. "Heading back to the cargo bay now."

"Nice one, Knuckles."

"OK, Pete, closing in on that visual, standby."

"Roger that. Nup, not a body, I think the crew were cooking up some sort of spit roast when all hell broke loose." He paused for a second before adding, "Scrub that, definitely a body. That's really... eww... my bad."

Pete scanned his displays again. "Krud. Coops can you tell Gandalf to remove corpse from the proximity warnings? I can't go a minute without getting side-tracked."

"Did you hear that Gandalf?"

"Yes, Cooper, system updated."

"Thanks. Can you get Bushaii Bartton on comms?"

"Unable to connect at this time."

"What? Why? Is he OK?"

"That is unclear."

"Dammit. Ginge, if you're still in visual can you skip ahead and check the status of the Gaujillo. Gandalf, keep trying to reach Bushaii. Let me know when you do."

"One step ahead of you Captain. She's in pretty decent shape – bridge looks intact. Oh, hang on; they've got some serious fire on the underside there. That does not look good. It's spreading quickly too."

"Thanks. Gandalf can you open up comms with the Gaujillo?"

"It is done."

"Attention crew of the Gaujillo, this is Captain Cooper Simpson of the UNS Dorsano. We're here to rescue you."

There was a slight delay before a response came – nervous but not panicked. "Captain Cooper Simpson? The Cooper Simpson? This is Er Otoh, High Commander of the Gaujillo. We are most honoured"

Cooper wasn't sure how to respond so kept his course set on the conversation at hand. "Can you advise on survivor numbers, High Commander?"

"There are 41 here on the bridge but the bulk of the crew are fighting to save the ship's core."

"Acknowledged. We are about to clear the…" Cooper scanned the UT to for further information on the skeletony-Pisary ship. "Pertura and obtain a visual of your vessel. Can you call them back to the bridge for rescue?"

"That would not be advised, Captain Simpson. If the core is taken by fire the results would be catastrophic."

"More catastrophic than we are already dealing with?"

"We estimate none of the remaining souls in orbit would survive the ensuing blast. That is why all who are capable of reaching the surface have done so."

Cooper looked around the room, gauging expressions as he contemplated. "Can we offer assistance in saving the core?"

"Negative, it is a losing battle. We are only creating time for others to be saved."

"How long do you think we've got?"

"Current estimates are 90 minutes."

"And if you cease your attempts to save the core and move all crew to the bridge for rescue?"

The hologram of Er Otoh looked to his right, clearly in consultation with other crew. "Thirty minutes."

"Pete, how quick can you get us there?"

"You're not serious?"

"I haven't got time for this right now, Pete."

"Exactly!"

"From here to the Gaujillo – how long?"

Pete paused seemed to do some calculating in his head before responding. "Like, 25 minutes or something, then there's manoeuvring and docking – could be half an hour, maybe more."

Cooper nodded and though on his next move.

"Excuse me, sir," said Hoffman.

"Yes?"

"I think the pilot may have overestimated the time to dock with the Gaujillo, sir."

Cooper looked Pete up and down. "Did he?"

"I think so. He told me 11 minutes moments before you asked, sir."

Cooper gave Pete one last extended glare before speaking. "Gandalf, beyond the Gaujillo, how many survivors do you detect still in orbit?"

"There are nine aboard the Asterion – near the docking bay and another 17 floating amongst the wreckage. The rest have either been rescued, perished or already reached Baltwae's surface on the remaining craft or escape pods. You also asked to be updated on the status of Bushaii Bartton and his ship and crew – I can confirm their craft was destroyed. There are no survivors."

An aching screamed out from Cooper's chest, he dropped his head and closed his eyes as the news sunk in. It was all too much to comprehend. He buried the thought and returned his focus to the task at hand.

"Pete: Gaujillo now!" he said as he reached for his communicator. "Knuckles, you got room for nine more?"

"Erm... just."

"Can you head towards the cargo deck of the Asterion – there are a small number of survivors there."

"Roger."

"Great. We've also got a number of random survivors floating around – Gandalf has marked them on the bridge display. Send this data on to all the rescue teams and organise your ships to grab them."

"Sure thing."

"Oh, and you're on a time schedule. Like, tight. You need to be back on the Dorsano or at the surface within, say, 25 minutes."

"Erm... shit... OK, confirm."

"Knuckles, language," said Pete.

"Shut up Pete," said Cooper and Knuckles in unison.

Cooper looked at his communicator. "We're rendezvousing with the Gaujillo. I'll leave Dilania and Hank at your disposal. Sing out if you need anything, otherwise, out."

"Roger that. Out."

Cooper returned to his attention to the holocomm with Er Otoh. "High Commander. We're on our way, call your crew back to the bridge."

"Captain Simpson, please reconsider, you're risking too much to attempt a rescue."

"It is already decided."

"See, he would've understood," said Pete.

"Wow... just wow," said Cooper.

But the conversation petered out as the Dorsano rose above the Pertura and the injured Gaujillo appeared on her horizon amongst a jigsaw of dancing debris pieces large and small. Fire had now taken the back of the craft, obscuring the cargo area and thrust modules. The ship listed towards both the Pertura and Baltwae's surface – although the speed of descent meant the craft was in no danger of hitting either object before it exploded courtesy of the core.

"Captain Simpson of the Dorsano, this is Erranj Meld of Baltwae command."

"Erranj, we've got a situation up here."

"It is understood – we are monitoring your progress. Have you considered how you'll move clear of the Gaujillo's blast zone once you've rescued the crew?"

"I figured we'd warp clear of the blast and return once it has cleared."

"A sound plan, should you be clear of the larger parts of debris in time."

"We will."

In the background Pete sighed.

"There is something else," said Erranj.

"Yes."

"We believe a number of significant vessels are salvageable – the Corrax, Themmington and GX 124: Narran. Our evacuation has left us in no position to save them. We know time is limited but we would appreciate—"

"Look Errenj, we've got a fair bit happening here right now, you sure there's no chance you can send up your own people?"

"Not possible, there is also a situation developing at surface level."

"Situation?"

"One of the returning vessels has collided with the docking bay entrance, at this moment no craft can come and go."

"Wait, what? I've got eleven rescue craft and troop carriers out there right now. They're under orders to head for the surface if they can't return to the Dorsano on time."

"You will need to revise that instruction."

"Shit."

"Does anyone flippin' listen to me?" said Pete.

Cooper blocked his mind to the chaos on the bridge and the chaos in the skies around the Dorsano. He knew every decision would have consequences – he knew there was a chance blood would be on his hands from the decisions he made at this moment.

"Thanks Errenj, we will see what we can do."

"It is appreciated."

Cooper closed the holocomm connection and looked around the bridge. "Gandalf, can you highlight the Corrax, Themmington and GX 124: Narran in the skies?"

"It is done."

"Estimated travel times by ferry?"

"The Corrax: seven minutes; Themmington: 11 minutes; GX 124 Narran: 17 minutes."

"Estimated time until the core of the Gaujillo explodes?"

"Twenty six minutes."

Cooper turned to Oil Man's main guy. "Reynolds, do we have any more crews on standby – to ferry crew?"

"Affirmative, Captain, I can have three troop carriers ready to go ASAP."

Cooper nodded his approval and noted a look in Reynolds' eye he hadn't seen before. Perhaps the faint glimmer of approval for his teenage leader?

"OK everyone, you heard the situation. I need volunteers willing to head to the Corrax, Themmington and Narran. If you can fly a Starship it would be a bonus."

Dilania rose from her seat. "My captain's wishes are my wishes, always."

Before the words had left her mouth Hensaro was on his feet. "If you need pilots, you need me."

"Thank you both, anyone else?"

The Ginge stood up.

"Don't be stupid," said Pete. "You can't fly a Starship."

"Have a look around, do you see anyone else with credentials? Besides, if you can do it, I sure as shit can."

"Flipping heck Ginge!"

Beside Pete, Luca stood up. "You'll stand a better chance with me at your side."

Pete looked at him, betrayed.

Cooper nodded his approval. "I think given th—"

Behind the captain, Alan cleared his throat. Both he and Rhiannon were on their feet, volunteering their services. Cooper's heart sank when he saw Rhiannon, he shot her a look to reconsider but she ignored it and stood to attention. "At your service, Captain."

Again Cooper paused for a change of heart. None came. "Very well, Hoffman is with The Ginge, Rhiannon with Hensaro and Alan with Dilania."

"Yes," said Alan instinctively and far more aloud than intended.

Everyone looked at him.

"Get yourselves to the docking bay ASAP – hit those ships and get out of here. The Corrax is closest; I suggest the Hoffman and Ginge take that due to their inexperience, I will take advice on who to send t—"

"We shall take Narran," said Dilania. "It is the riskiest mission and I have the greatest experience."

"No," said Alan instinctively and far more aloud than intended.

***

CHAPTER 21

Pete nudged the right bumper on his controller a couple of times and The Dorsano eased in that direction. The hull of the Gaujillo hulked over the bridge like the face of an epic Himalayan mountain.

"Careful," said Cooper as he stared

"You're kidding me! This is a careful situation? Wow, thank god you're captain and can tell me these things."

"Sorry, I just..." said Cooper before realising no words could atone for the previous statement.

"Coops, we're here," came The Ginge's voice through the captain's communicator.

"Brilliant! What have we got..." asked Cooper to the air before answering the question himself. "Eleven minutes."

"Yeah, it took a little longer than we thought – it's pretty crazy out there."

"Are you able to stay put for a minute? I've told Knuckles the rescue craft can hitch a ride with you or the other if it's more convenient."

"All good, there's not much traffic above us – we can shoot up and out of here in no time."

"Roger that, great work, Ginge. Out."

"Out."

Cooper interfaced with his communicator. "Knuckles, you there?"

"Here, I'm also joined by nine new friends from the planet Azzhatt."

"Legend. You heading back to the Dorsano?"

"Negatory there Capitano. We're not far from the Corrax, heading that way. Besides, I hear it's the first time in history a redhead will be at the helm of a Starship – we can't miss that. Sorry."

"Roger that, Knuckles. No offense taken. Out." He turned to the military commander now multitasking in the absence of others. "Reynolds, update on the other craft."

"Hensaro and Rhiannon have made good time and are docking with the Themmington as we speak. Dilania and Woodcock are still five minutes away – on schedule though."

Cooper felt a wave of relief hit him; he tried not to show it. "Nice one. And the rescue craft?"

"Close. Still three people to pick up but everything's on target. Everyone's being redirected to the closest starship – not sending anyone to the Narran though – just no window."

"Great work, Reynolds."

The Dorsano was approaching the Gaujillo along its spine, the craft back-to-back to ensure the two bridges were as close as possible for the crew transfer. Soon the bridge of the Gaujillo came into view for Pete, Cooper, Hank and the skeleton command crew aboard the Dorsano.

Pete gave the left trigger a quick dab, then another as the reverse thrusters killed much of the Dorsano's momentum.

"I can see them," said Cooper. He saluted and received what he assumed was a greeting in return – Cooper found it a surreal exchange.

Pete looked up to see the surviving crew of the Gaujillo looking up at him from their bridge. "Oh my God, I didn't think there'd be that many. How long's that gonna take?"

"You worry about lining up the hatches, I'll worry about the transfer. ETA?"

"Um, almost there," said Pete as he tapped the left trigger again.

Soon the hatches were in alignment, to the triumphant shouts of bingo from Pete. A walkway was deployed from around the Dorsano's hatch to envelop that of the Gaujillo.

"Pressurising," said Pete. "Five, four, three... and we're good to go."

Cooper looked at Er Otoh's hologram. "You are good to commence boarding high commander. But we don't have much time."

"Acknowledged," said Er Otoh. "Transfer commencing."

"Nice work," said Cooper as he slapped Pete on the back.

"Ouch," said Pete.

"Reynolds, how are the others going?"

"Dilania and Alan are approaching the Narran, all known survivors have been rescued and the last three craft are heading for starships – one rescue craft headed for the Themmington, a rescue craft and troop carrier heading here. No known delays, all within the margin of error."

"Good. Keep me posted."

"Yes, sir."

"Gandalf, how are we looking?"

"With your eyes if my vast knowledge of human biology is accurate."

Cooper rolled said organs. "With the crew transfer."

"Estimate three minutes and twenty seconds until completion."

"And time to the core explodes?"

"Five minutes forty. Estimated."

"Pete can you flip this thing up and out of here in a couple of minutes?"

"In theory."

Cooper ignored the comment and slapped him on the back for encouragement. "Ouch."

Hensaro's hologram appeared in the centre of the bridge. "Captain Simpson, we are aboard the Themmington."

"Brilliant. You've got a rescue craft headed your way, once it lands you are free to blast out of here. Out."

"Acknowledged, captain."

Cooper reconnected with The Ginge on his communicator. "Ginge, my friend, you are good to go."

"Really? Can't we help or anything?"

"Yes, by going. It's time we got some runs on the board here."

"Maybe we cou—"

"Leave. Now. Out."

Just as Cooper closed the link a shudder rumbled through the bridge. He looked at Pete. "What was that?"

"I have no idea."

"Reynolds – anything?"

"Negative."

"Flapjacks!" squealed Pete as he started manically adjusting the controller.

"Flapjacks?" said Cooper.

"Something's wrong."

"As in 'not right' levels of wrong?"

"I think so – the Gaujillo – something's happened," said Pete as he feverishly manipulated his controller so the Dorsano could match the Gaujillo's change of direction.

Cooper looked up to see the crew of the other vessel in a panic to get to the walkway.

"Er Otoh – what's happening?"

"There was an explosion, telemetry down, unable to assess."

They stared at each other across the bridge divide. "The core?"

"I don't believe so, we are still alive."

"We're still linked, get your people across."

"Affirmative, but if it becomes unsafe, you are to save yourselves, understood?"

"It won't come to tha—"

Another explosion shuddered the Dorsano, this one bigger than the last. Cooper looked up at the crew of the Gaujillo, most had been knocked from their feet by the blast.

Then the Gaujillo started moving up and away from the Dorsano at the stern.

"Tantrums!" cussed Pete at the controls.

"I preferred flapjacks, to be honest."

"It's moving too quick, Coops, I can't stay with it."

"Hang tight, Pete, another minute or so and we should have them all across."

But even as he uttered the words, Cooper looked up to see the crew of the Gaujillo in disarray – injured strewn across the floor and only a handful entering the transit chute. "Hang in there, hang in there."

Pete manipulated the controls and sent the Dorsano into a rotation in an attempt to match the Gaujillo's new trajectory. He tried to equal the movement without overdoing it, knowing too much thrust would send the two ships into each other. Despite his best efforts, the ships were separating, the transit chute seemingly stretched to the limit.

A third explosion ripped through the Gaujillo. This blast so massive the vessel both shot away from and pushed backed the Dorsano in turn. The walkway held the two ships together a brief second longer before tearing from the Dorsano's hatch. Bodies and debris were sent into the cold vacuum of space.

Cooper looked up to see the faces of Er Otoh and his crew screaming in terror as their fate became clear. Around him, warning indicators started flashing in worrying numbers – system failures, hull breaches and proximity warnings.

"We've been hit," said Reynolds.

"Damage?"

There was a pause while Hank tried to absorb as much information as possible in a short time. "Lots."

"Great! Pete, can you still fly this thing?"

Pete toyed with a few of the basic controls. "Think so. Ready to blast out of here?"

"Any chance you can make a pass by the Gaujillo?"

"We've got less than two minutes as it is!"

"Can you get us close?"

"No point, walkway's gone," said Pete as he starting to get the Dorsano's movement under control.

"If we can cruise close by they may be able to teleport."

"Isn't that dangerous?"

"Yes, but it's their only option."

Pete had the ship facing towards the Gaujillo. "I meant for us, you know, with it blowing up and all."

Cooper looked out the viewshield at the Gaujillo, far enough away to see the damage from the explosions. The damage was vast – fire engulfed the majority of the ship now and a huge crater spewed loose pieces of craft into the void.

"High Commander, are you there?"

"Yes, but why are you?"

Cooper noted how calm the Er Otoh was amongst the chaos around him, it made him more determined to help. "We're going to do a flyby on our way out of here, it should give your crew a chance to teleport."

"It's too dangerous, save yourselves!"

"Estimated time of closest approach..." said Cooper as he looked at Pete.

"Thirty two seconds," said Pete as he punched the Dorsano forward.

"Thirty two seconds. Prepare your crew."

"Affirmative... and thanks."

Then another explosion rang out, near where the large crater was, it ripped the back of the ship apart and sent the front of the vessel flying off at an angle. The new trajectory of the bridge sent it on a collision course with another piece of wreckage resulting in another explosion.

The Gaujillo was lost.

"Nooooo!" cried Cooper as the alert intensity in the Dorsano's bridge intensified again.

"It's heading right for us!" said Pete as he jostled the controls to send the Dorsano into a barrel roll.

"Proximity alert: catastrophic," said Reynolds.

"Yeah, I got that," said Pete before adding. "Shivers, blast, turds!"

"Fifteen seconds to impact," said Reynolds.

"Forty seven seconds to core detonation!" said Cooper.

"Dung, snot, armpits, poop!"

Cooper and Reynolds rode Pete's every move and PG-rated cuss as the ship rolled over and up.

"Impact in ten, nine..."

"We're not gonna make it," said Pete.

"Six, five..."

"You can do it, Pete."

"Three, two, one..."

Pete screamed. Cooper screamed. Reynolds screamed.

A large metallic ripping sound lurched out through the bridge and through the bodies of the crew. It was long and loud – not even the additional alert tones could drown it out.

After what seemed like an age, the groaning stopped.

"Pete, open your eyes," said Cooper as he turned to Reynolds. "Damage?"

"Everywhere."

"Shit!"

"Coops!"

"Really Pete, even now? Just aim for the biggest gap you can find. Gandalf, are we capable of hyperspace?"

"There are too many variables to calculate."

"Not like we have a choice. Eighteen seconds to detonation. Pete, find a gap and punch it."

Cooper and Pete exchanges glances and prayers in an instant.

Pete refocussed on the game of spaceship carcass asteroids at hand, bashed away at the controller until he was content. "Got it, I think."

"You think?"

"Just don't blame me if we die!"

"I'll try not to. Hit it!"

Pete screamed as he hit the button, they all did.

***

CHAPTER 22

"Where has all the wreckage gone?" said Luca Hoffman.

"Search me," said Ginge. "Lilly, you there?"

"I am everywhere," replied his avatar to the UT.

"I bet you are," Ginge added with a smile. "Listen, toots, where is everyone? And, well, everything?"

"That data is unavailable at this time."

"What do you mean unavailable?"

"The core explosion has left a number of systems well under maximum capability. Communication functions beyond the Corrax will be offline for several minutes."

"Wow, must've been one hell of an explosion there, honey. Any chance you can update us on the other ships."

"That data is unavailable at this time."

"Ah yeah, of course," said The Ginge, trying not to miss a beat in front of Hoffman. "Hey, babes, can you open up comms to Baltwae"

"That function is unavailable at this time."

"What? OK, yeah, sure." The Ginge looked at Hoffman. "She's not just a pretty face."

"Wait," said Hoffman as the pieces of the puzzle started to fall into place. "Is that Lilly as in Lilly Longmire?"

"It is."

"Five time AVN winning actress?"

"The first woman to be able to take a full—"

"She's standing right there, show some class," snapped The Ginge.

"This from the guys who calls her toots?"

"That's... different... Lilly and I go way back."

Hoffman looked him up and down. "Sure."

After an extended silence the two sought different time-killing activities at opposite ends of the Corrax's bridge. Luca studied his navigating display, despite the fact most functions weren't working, while The Ginge stood at the viewshield, waiting for one of the other ships to reappear.

It wasn't long before The Ginge saw movement in orbit. The stars shimmered, blurred then warped in peculiar ways before a small blue glowing light appeared at the centre of the phenomenon. It grew slowly for a second or two before pulsing aggressively in a flash. When the bright light receded a starship was floating in its place.

"Got one," said The Ginge.

Hoffman left his console to join The Ginge at the viewshield. "Who is it?"

"Not sure, Themmington or the Narran, hopefully. Lilly, can you identify that ship?"

"That vessel is the Themmington."

"Cool, can you open comms?"

"That function is unavailable at this—"

"How long is this going to go on for?"

"That data is unavailable at this time."

"Seriously? You're annoying, anyone ever tell you that?"

"The Universal Translator gets called a wide variety of names, annoying is amongst them."

The Ginge had a sudden wave of guilt. "I'm sorry sweet cheeks, it's not your fault."

A digitally muffled noise broke out behind The Ginge and Hoffman, they turned to see Rhiannon's hologramatic image flicker into and out of existence. They ran over to face where her likeness had appeared.

"Rhi!" said The Ginge. "Lilly, can you holo me back to the Themmington?"

"It is done."

Rhiannon's signal came through again, stronger than before. "...ou there? Ginge? Is that y..."

"Rhi! Yes, can you hear me?"

"Yes, Ginge, thank God," said Rhiannon before she turned her head. "I've got him."

"You made it," said The Ginge. "I mean, obviously. You all OK?"

"...or... aching..."

"Say again Rhi, you're breaking up."

There was a digitised sound as Rhiannon's image dropped in and out of view. "...up... repeat, you're breaking u..."

"Lilly, how long's this going to go on for?"

"Connections are being re-established, full local communications should be available shortly, contact with Baltwae and further afield will follow shortly."

The Ginge was about to thank his avatar to the UT but was interrupted by Rhiannon's hologramatic likeness, with a signal stronger than before.

"Repeat: Are you reading me now?"

"Loud and clear, Rhi. You guys all good?"

"Yes, you?"

"Still here, we've got a couple crews in the dock as well as four survivors. Total: thirty three."

"Seventeen here," said Rhiannon. "Where are the others?"

"We've got nothing, we were flying blind until you rocked up."

Rhiannon started to speak again but her signal broke up, seconds later a disc of space warped nearby as a third craft settled into orbit – the GX 124 Narran.

*

"Where the hell are they?" asked Knuckles as the time clicked over to one earth hour since the explosion.

"They're obviously just sorting a few things out before they come back," said Alan in a tone that suggested even he wasn't convinced of his words.

"So, at what point should we start being concerned?" said The Ginge to both the other crews on the bridge of the Corrax and those in hologram form from the Themmington and the Narran.

"About forty minutes ago," said Rhiannon.

"We can't just stay here, it's doing my head in," said Knuckles. "Let's head surface-side."

The Ginge nodded. "Good call, seconded."

There was a pause as everyone waited for the idea to become a directive. "Um, who's in charge?" asked The Ginge eventually.

"What does your command structure dictate?" said Hensaro.

The Ginge looked around at the other humans from the command deck – silence was the collective response.

"You do have a command structure."

"Oh, yes, totally, I think. Maybe," said The Ginge as he looked at the other humans and pleaded for someone else to take the reins of the conversation. There were no takers. "Let's just say, for example, we had a command structure, thingy and we... lost it."

"You lost it?"

"Yeah, let's just say it was on a piece of paper somewhere – like totally official paper – letterhead and everything..."

"What is letterhead?"

"...but that paper was on board the very ship where most of the senior people on the command structure actually were."

"Yes?"

"Well, if we lost the commanders and the little piece of glossy, 140 gsm paper with, like, pretty little full colour diagrams and everything – in that totally unlikely example, who might be the next in charge?"

Hensaro stared at The Ginge over several seconds before responding in what The Ginge correctly felt was a patronising tone. "You don't have a command structure, do you?"

The Ginge didn't respond with words. He didn't have to, the expression on his face didn't require the universal translator to be understood by all in the conversation.

"Well, in that hypothetical circumstance, where your captain, military commander, chief tactician and senior pilot, along with the actual command structure itself were lost..." said Hensaro.

The Ginge nodded intently, as if doing so would officially erase all knowledge of the previous revelation.

"...in that circumstance I believe senior tactician would be in charge."

The Ginge, who had continued nodding throughout Hensaro's words, finally stopped when he realised that mean him.

"What? What about you or Dilania?"

"While we are more experienced, this is a human mission."

"What about..." said The Ginge, as he looked around the remaining humans. "OK then, me it is. Everyone prepare your crews to land on Baltwae. We'll see if they know anything we don't."

"Hang on, wait up, what about me?" said Knuckles.

"What about you?" said The Ginge.

"As acting captain thing, butt-munch. I saw you look at me right now, didn't even think I was a possibility."

"Dude, you're the gunner."

"And?"

"Your job aboard this ship is to shoot at enemy spaceships."

"So? What makes senior tactician any better?"

"Probably the senior bit... and the tactician bit."

Knuckles thought on which stinging retort to bounce back at the wannabe acting captain before settling on. "Shut up."

There was a decidedly uneasy silence from the rest of the beings in the room. Eventually The Ginge realised the only way to stop it was to start doing his acting captain thing. "Alright people, you heard, we head to Baltwae. Catch you on the other side."

*

There was something triumphant about making landfall after a long or turbulent mission. Each of the 84 beings who stepped foot/claw/paw/ undercarriage on the docking bay felt it in their own way. For the survivors, rescued from fractured craft or the void of space, the feeling was overwhelming – some cried, some prayed, some celebrated, some embraced their rescuers, some sang. The rescuers, too, shared similar emotion, though not as extreme.

Despite the cold alien world and the lighter-than-Earth gravity, it was as if the physical act of placing feet on solid ground somehow sucked the anxiety away. It was one step, one act, one moment that said everything was going to be fine.

Once each realised their survival was assured, thoughts immediately turned to the many who weren't so lucky. The makeshift crews of the three starships intermingled in front of the troop carriers and rescue craft that had brought them to land. Crew offered each other consoling hugs and actions replaced words.

The docking bay was vast. The Ginge, his remaining crew and the rescued survivors, found themselves at the far end of one wing of the bay. Medium and small craft of endless variety stretched all the way back to the central hub and the entry portal, still charred and damaged by the crash only an hour or so previous.

Rhiannon ran over to The Ginge and embraced him. "I'm scared," she whispered eventually.

"Me too," said The Ginge. "But he's gonna be back."

She took a step back, nodded and wiped away some tears.

"He is alive," said Dilania as she moved in to greet them. "I can feel it."

The Ginge embraced her, as did Rhiannon, who even managed a slight smile at Dilania's certainty.

"Don't forget your Uncle Alan," said the former sci-fi convention MC before positioning himself in hug mode.

Rhiannon withdrew her arms and shuddered.

"What did I say?"

"I don't think the whole Uncle Alan thing helped," said The Ginge.

"Too familiar?"

"Too creepy."

Alan hunched. It was enough make Rhiannon laugh through pity. "Oh come here, you idiot."

Alan followed his DNA-driven protocol to obey everything a woman says, embracing Rhiannon.

"But if you ever call yourself Uncle Alan again I will personally kick your arse."

He smiled. "Point taken."

"Ladies!" said Knuckles as he slapped The Ginge hard on the back to remind him who was stronger, a small victory after the acting-captaincy defeat.

"Here he is," said The Ginge as he offered up a high five handshake trying not to show the pain the ripped through his back.

"It is good to see you again, Captain," said Hensaro as he joined the gathering.

The Ginge nodded. "Get in here, buddy."

"I'm not sure what you mean."

"C'mon, share the love," said The Ginge as he opened his arms and welcomed Hensaro.

"The love?"

"Just a hug, Hensaro. Man love – no big thing."

"Man love?" said Hensaro as he took a step back.

"Not that kind of man love."

"I am not sure what you want from me, Captain. Or whether I have the inclination to give it."

"Just get over here."

"Yes, my Captain."

Hensaro swallowed nervously as he approached The Ginge. When he was within range, the acting captain threw his arms around the Fredahn. To ease the tension of the observing crowd The Ginge put his weight on his right leg, then kicked up his left leg, pointing his toes skyward.

Rhiannon laughed, Knuckles called him a tool and Uncle Alan nodded knowingly.

"Captain, a welcome party approaches," said Hoffman.

The Ginge let Hensaro alone then he and his crew looked up to see a fleet of hovers, dwarfed by the troop carriers and rescue vessels, heading their way.

When within a couple of lengths, the dozen or so vehicles slowed to a halt, several beings from a variety of species hopped to the floor and approached.

"Fleet Admiral Hensaro Althot, Grand Captain Dilania Duritee, Acting Captain Wesley Ellis—"

"People call me The Ginge."

"...The Ginge and crew of the Dorsano; I am Erranj Meld, what you have done today will not be forgotten by my people. You have saved many lives and ships from the ashes of total destruction."

"We just did what we could," said The Ginge, instinctively striking a more heroic pose and nodding to a couple of females from the greeting party.

"To have the legendary humans from Earth in the same place as the second highest ranking Fredahn and our people's greatest warrior ally is truly an honour beyond words."

The Ginge did a double-take and looked at Hensaro and Dilania through a new lens. He rifled through memories of his interactions with them, hoping nothing too embarrassing emerged. But all recollections were blocked by the recent man love incident and his heroic stance abated somewhat.

"Erranj, what word of Bushaii Bartton?" said Hensaro.

"It is my grave duty to report the Vexor was destroyed. All aboard were lost."

Hensaro stepped back and made a gesture with his hands and face which was unfamiliar to the humans. "He died great and will be remembered great."

Behind him Dilania repeated the phrase as did the greeting party.

The Ginge lead the humans in acknowledgement but also had other things on his mind. "Any news on the Dorsano?"

"We have no word, the UT is still repairing and we have no communications beyond this planet yet."

The Ginge attempted to remain composed while he nodded to Erranj. "Any estimates on when communications will be restored?"

"It will still be several hours before full capacity is reached."

The Ginge nodded.

"In the meantime, I've announced, Haasark," said Erranj before looking at Hensaro. "If our new leader chooses to endorse my decision."

"Haasark is a sage suggestion, my friend. The only suggestion," said Hensaro.

"Haasark?" said The Ginge.

"Tonight we celebrate the lost and we celebrate not being amongst them."

Hensaro started towards the hovers before turning to face The Ginge. "Come, you may even be able to find some kindred souls to perform this man love pastime with."

"What? No! I—"

"Don't fight it, Ginge," said Knuckles.

"I..."

***

CHAPTER 23

The void above Baltwae warped once again and a forth ship zwipped into orbit. The Dorsano, worse for wear, but functioning limped into position alongside the Themmington.

"We still in one piece Pete?"

After a quick consultation with the floating digital interface he gave his captain the answer they both already knew would be the case. "All systems operational, Captain."

Hoots and applause from the skeleton staff echoed around the bridge.

Cooper smiled and looked out the viewshield, the subtle exterior lighting of the three other starships dominated the view. He closed his eyes briefly and let the relief run over him – they were alive, Rhiannon was alive. It was only then he let the emotion get to him. "Well played everyone, that was unbelievable!"

He walked to Pete and hugged him, then completed a lap of the room, shaking hands and praising skills. He ended in front of Reynolds and offered out his hand. Reynolds sat in observation of his captain for a moment, saluted then offered his hand.

"You did well, Captain," said Reynolds.

Cooper nodded at him before turning to the chief tactician. "Rodriguez, open comms with Baltwae, I believe they'll be expecting us."

"Yes, sir."

"Commander Reynolds, prepare your men to shuttle the crew surfaceside."

"Yes, sir."

"Everyone ready for some R and R?"

"Yes, sir," came the collective response.

"Captain," said Rodriguez, "we're not getting a response from Baltwae."

"What do mean?"

"I've used comms protocol with the command centre and the docking bay – nothing."

"Gandalf, are you there?"

"I am everywhere."

"Is everything OK on Baltwae?"

"Define OK."

"Well, between the operational staff and the survivors there should be a couple thousand people down there but we're not getting any contact."

"There are 3283 beings currently on Baltwae."

"So, where are they?"

"They are observing Haasark in the great hall."

"Haasark?"

*

"I have been a big admirer of yours since Galactica, The Ginge. If you wish to spend any one-on-one relationship time with me tonight I would make it worth your while."

The Ginge looked the person behind the offer up and down – a humanoid with an elongated nose, flowing dark hair and striking purple eyes. "Look, thanks for thinking of me but, like I told the others, my bread isn't buttered on that side."

"I'm not sure I understand."

"I don't bat for the home team?"

The being, known as Gwerth Ballar, stared blankly at him, then took another swig of his drink.

"The thing about man love—"

"Man love, yes."

"Man love, no. It's just a saying, on Earth. It was not meant to be taken literally... by, like, so many people."

Gwerth finished his drink in one swift movement, then delivered The Ginge a disapproving glare before moving on. The Ginge let out a groan then indulged in some of his beverage.

"You know Ginge," said Knuckles, "I've always been a little jealous of the way you attract potential alien lovers like moths to a flame, but not tonight for some reason."

"Not now, Knuckles."

"It's, like, now you're captain you've taken things to another level."

"Can we just talk about something else for five minutes?"

"Yeah, another level. You've gone from chick magnet to... dick magnet."

"Really? Try to insult me, fair enough, but could you aim for something a little more sophisticated than dick magnet?"

To add injury to insult The Ginge felt a stinging ping up form the back of his head, as said body part was thrust forward. He realised someone had slapped him and turned to see Cooper standing over his seat, he did not look pleased. Behind him Pete was grinning in anticipation.

"So, let me get this straight," said Cooper. "Your starship, your captain and about 95% of your crew were missing and you decide it's a good time to go for a drink with the lads."

The Ginge jumped to his feet and went to salute before realising he had a beer in his right hand. He swapped the beverage to his left and acknowledged his leader. "I... I..."

"Tell me about it," said Knuckles as he greeted Cooper and Pete. "It's been like the butcher's section of the supermarket here. Everyone's taken a number, hoping for ginger sausa—"

"Shut it, Knuckles," said The Ginge.

Cooper stared at the pair blankly.

"Man love," whispered Knuckles.

"Shut up!" spat The Ginge.

"I don't really know what you're talking about and I don't really care. I'm slightly more concerned with why you're here getting tanked while the ship was missing," said Cooper. "We lost people up there, you know."

The Ginge felt compelled to lower his drink to the table. "I know; we know what happened. And it's not like that Coops. When the UT was strong enough to monitor your status, it was the first thing we did, right Knuckles?"

"That is true."

"Then when we saw you were alive and running repairs to get the ship ready to handle another hyper jump, well, we celebrated long and hard. In fact we're still going."

"Didn't think to send out a greeting party for when we got back?" said Cooper.

"Absolutely! I just... forgot."

"C'mon Coops, relax," said Knuckles as he offered up a fresh beer to Cooper and Pete. "It's all good now. Besides, to be fair to the acting captain—"

Pete sputtered his first mouthful of ale. "Acting captain?"

Cooper stared at his redheaded friend long and hard, much to the delight of Knuckles and Pete. At some point in their eye contact The Ginge smiled and Cooper's stern look broke. They both laughed.

Knuckles studied the two, feeling a step or two behind what was going on between the best mates. "So, what, we're good now?"

Cooper and The Ginge laughed then clinked glasses.

"Yeah, we're good," said Cooper.

"Cooper!" cried Rhiannon from across the room as she sprinted towards him.

Cooper faced her, offering a large grin and open arms. She closed the ground in no time before leaping onto him. They kissed, while the others stood around awkwardly.

"Get a room, you two," said Knuckles, eventually.

It was enough to prise the two apart, but they were still too wrapped in other's company to show any signs of guilt.

The other three formed their own circle of conversation. Pete looked at The Ginge. "Acting captain, wow, you nailed that."

"Shut up Pete, I was just bringing my own thing to it."

"It's true, he was," said Knuckles. "Never have I seen a captain get so much male attention – and I include Dilania in that."

Pete looked at The Ginge with a 'please explain' expression.

"It's just a... miscommunication. Look, can we talk about something else?"

"What about the fact you managed to spend your entire captaincy in various stages of drunkenness and without a ship," said Pete.

The Ginge tore through his mind for a witty comeback, when none presented he raised his glass. "I'll drink to that," he said before offering cheers to the others.

The boys clinked glasses just as a beaming Cooper and Rhiannon joined them.

"To war," announced Cooper.

"And Quel," added Pete while looking at Knuckles.

They laughed then drank, to either distract their thoughts from recent events or to celebrate that they were all still around to do so. Behind the smiles they each felt the shadow lurking, one which told them life was cheaper in this new world and moments of loss may well be the norm. It was only then they started to understand Haasark.

They smiled at each other again. They were back in the black, no idea what they were doing, in over their heads and loving it.

*

Cooper watched the replay of the Baltwae massacre for a third consecutive time. The presentation had been put together by Erranj Meld, who was delivering it to the senior crew of the Dorsano as well as the remaining key players from the other ships. Meld's breakdown of how the battle unfolded was thorough and insightful. It left Cooper feeling almost totally helpless. They had been outnumbered and outgunned, overwhelmed by a superior force. There had been no warning and, worse still, their supposed secret location had been compromised.

As Meld's presentation concluded it was clear Cooper's helpless feeling was shared by all. The room fell silent and all eyes fell on Hensaro for direction.

"Comprehensive, Erranj, congratulations," he said. "The question is where to from here."

Dilania, who was sitting to Hensaro's left, didn't miss a beat before responding. "We still have four starships, we rebuild and rejoin the war effort as soon as possible."

"There are no shortage of fleets in the local systems," said Erranj. "Or we could join the forces at the Tillaron front, tactically that's a vital region to hold right now and they have suffered many losses."

"In a similar manner to what happened here, I understand," said Hensaro.

"As you say."

Hensaro pondered the information as he scanned the faces in the room. "Captain Simpson. It would be nice to hear from our human comrades, after all, many of us would not be here without you."

"What? Um, yeah, sure," said Cooper stalling for time while he thought of something to say. "I'm just wondering why we don't spend more time working out why and how the attack happened here."

"Yes," said Hensaro, in a tone that encouraged more.

"Well, this was supposed to be a secret location. How did they even know we were here? Maybe, finding out that answer could help us predict what they will do next."

A tall, thin leathery creature, with speckled warty regions, stood at the far side of the room. "Are you suggesting we have an informer in our midst?"

"I... I'm not sure, maybe?"

"That's outrageous. An accusation like that can split New Council apart."

"Kace Astoy, sit down," said Hensaro. "We seek understanding, not posturing."

Astoy glared at Cooper before returning to his seat, Hensaro nodded toward Cooper to continue.

"Yeah, so," Cooper started before realising he wasn't sure where to go next. He had ideas but should he raise them here?

Hensaro read the situation in an instant. "Captain Simpson, you are here for your opinion, if I wanted to hear from people who all think the same I would ask Kace Astoy the same question 27 times."

There were murmurs of laughter. Hensaro glared at Astoy before continuing. "It is also interesting, young human, some people have forgotten that they would still be part of the artwork on Galactica, working for our enemy, if you hadn't chosen a different path to others. History and allegiance are important.

"Tell us," urged Hensaro, "what would you do?"

"Ahh, well, I guess I'd try to find out how they knew about this place. Is there any way we can trace their movement – where did they come from, where did they go?"

"All good ideas, what about us? What can four starships in the middle of nowhere do to help New Council?"

"Well, I don't know about the other species, but we need training. I don't think we'll be much help to anyone until we have some more simulator hours under our belts."

"Yes, training is essential. Not just for you humans, all of us are facing challenges beyond our experience. We need to work together to advance our skills, just as Bushaii Bartton had intended – if on a smaller scale."

"Cool. Besides, Gandalf tells me it'll take some time to restore the starships to complete working order."

"Gandalf?"

"Sorry, the UT."

"That is correct," said Hensaro. "So, we research our enemy, possible information leaks and train our forces while our ships are repaired to full functionality."

Cooper nodded. "Yeah."

"Then what?"

Cooper looked around the room acutely aware of judgmental eyes upon him. "That would depend on what we discover, and how we train."

"Yes," said Hensaro, probing.

"There are too many outcomes to narrow down a guess."

"Thank you Cooper Simpson. That is exactly the kind of thinking we need. We do not simply react to what happens, we learn, understand, improve and plan."

Again, Hensaro studied the faces of those in the room as if he was glimpsing into their souls – such was the power of his stare. "You did make one omission in your observations, Captain Simpson. That is our greatest asset."

Cooper looked on intently, lost in Hensaro's natural assurance.

"Four starships – we travel light and more likely to go undetected. While the answer may be to align with a larger fleet, if that's where our training and intelligence leads us, we may, in fact, find the disaster inflicted on us may make us far more destructive than our enemy could imagine."

Despite Hensaro's conviction, it was clear to Cooper that many other didn't understand his previous statement, one which Cooper wasn't entirely sure he embraced either.

"But that is a discussion for another day. Today we acknowledge heroics and celebrate, tomorrow we begin the task of rebuilding our fleet and our strength."

*

Just over 3000 people attended the ceremony, it made for an overwhelming and odd collection of eyes upon Cooper and the crew of the Dorsano. The event was moved to the docking bay when they realised the grand hall, deep in the heart of the cold bedrock of Baltwae, would not accommodate the crowd.

Erranj Meld and his people had adorned the epic space with traditional Fredahn colours, fabrics and motifs. Behind the sizable stage, where the entire crew of the Dorsano now sat, stood twenty six large pedestals, each with a 3D planet rotating above them – symbolising the species that were gathered here together for a war most never saw. Between the design of the space, the people and the event it was easy to forget this area was glorified spaceship hangar.

The last hour had been a series of accolades to the crew of the Dorsano as the survivors of the massacre each stepped forward. In turn, each stated their name, rank, ship and species before adorning one of their saviours with a gift from their species and a narrow piece of decorative metallic trim for their uniform. The trim, in a silver looking finish, was worn around the base of the collar, the soft material substance was emblazoned with fine glyphs which, translated from Fredahn read: To save a life is to live your days with two.

Most of the survivors, some who'd had significant injuries not a day ago, were now completely healthy, thanks to the UT.

Cooper had to wipe a tear away from his eye on more than one occasion. Whether from the responsibility he felt for the two human crew killed returning to the Dorsano after rescuing the injured, the flashbacks he had seeing the terror Er Otoh and his crew in their final seconds, or the bravery they showed in the minutes leading up to it. The regret and pain also matched with pride for the way he and his crew had helped so many others. The emotions collided, threatening to overwhelm him into further bouts of tears.

After the parade of survivors had finished, Hensaro Althot moved from his position amongst the Dorsano's crew to a position near Erranj Meld, they were joined by another being – Ulnor Re Kotun. Cooper recognised the shimmering blue skin as that of the Kindria, the grey variant species that populated the ill-fated Gaujillo. Again moments flashed by him.

Erranj Meld stepped forward to address the crowd. "This place, Baltwae, has been a long time in the making. When my people discovered it floating anonymously through the void, its value in secrecy was immediately obvious. This place was an unusual asset in a war where difference, real difference, is rare. I didn't hesitate for a moment to be part of building, training and planning the offensive that would be launched from here. In truth, I felt, no believed, that this fleet would do something special. It seemed like destiny to me that the war would be changed from what we did here."

He laughed, a sarcastic chuckle filled with bitter irony. "Who knows, that may still happen. Whatever the case, I have witnessed heroics I will never forget. Acts of bravery I'll spend the rest of my life talking about. Captain Cooper Simpson, can you please step forward to join us?"

Cooper rose from his seat slowly then crossed the stage to the various sounds of appreciation the crowd generated. He felt thoroughly awkward as if the acknowledgement for him far outweighed his efforts. His feats, if anything at all, certainly didn't warrant this level of reaction, nor singling out from his crew. All he'd done was think of it like a puzzle, each step he took was the most logical response to keep as many people alive as possible in the circumstances presented.

He was soon joined by the Erranj, Hensaro and Ulnor and before he could think how best to handle the first greeting, Ulnor performed what must have been a traditional greeting or acknowledgement on him, then presented him with a small package. Cooper looked at the gift as Ulnor indicated to open it.

It was another piece of metallic trim, again in silver, but this one was not a long thin line to position down the collar, this was a symbol, a winged animal with glyphs and other adornments.

Cooper stared at the remarkable piece, then Ulnor, then his crew. He did not feel comfortable with where this was going.

Ulnor noticed Cooper was unsure what to do. He picked up the piece and spoke to the crowd and Cooper as he set about positioning the object on Cooper's chest. "This is the Jajawera Insignia. The Jajawera is native to my home world and its traits are known to Kindria throughout the galaxy. The Jajawera has possibly the most succulent meat on Falcray, but it is rare to taste. When a Jajawera is injured or in trouble others in the immediate area will come to its aid. They will put their life at risk to save another – for no other reason than to save another.

"So passionate and complete is this act it is rare a Jajawera is left behind. This insignia is one of the highest honours of my people. You cannot pretend, you cannot bribe, you cannot cheat to obtain this. This insignia is an echo of the soul and it is a rare blessing. What Captain Simpson and his crew risked to save 96 of my people, when the easy option was self-preservation, that cannot be measured.

"But it can be rewarded, Captain Cooper Simpson of the UNS Dorsano from Earth, I hereby award you the Jajawera Insignia."

Ulnor positioned the piece on the right shoulder of Cooper's uniform.

"Um, thanks Ulnor and everyone, it's a great honour. In the tradition of the Jajawera I accept this on behalf of the entire crew of the Dorsano. And to everyone else here, if we continue to have each other's backs we might still become a powerful force yet."

***

CHAPTER 24

Four months later.

The Ginge stood in the centre of a large, featureless space, all around him a hologramatic space battle played out. This was more than the fully immersive games he had created on Galactica, this was a war game on an epic scale.

As the lights of battle danced around the room, showering him in a night-club of radiance, he smiled. He was watching the attackers of the Baltwae force – humans, Fredahn, Jarden, Kindria and everyone else – coordinate a large-scale dogfight against an enemy force of his making. Real attackers, computer generated enemy, full scale war.

The smile on his face broadened as he thought back to the humble beginnings of his game-creation mastery with the new intake newbs on Galactica.

Around him, Sam and his camera crew jockeyed for position, taking shots of the The Ginge and his lightshow spectacular from numerous angles. They were barely a distraction for The Ginge anymore, merely a reminder to focus on his posture and keep his chin up.

"Here it comes, here it comes," he said as he rushed through the display of the dogfight, drawn in by the enemy starship and the Batlwae wave attacking her.

The camera crew rallied in an attempt to keep The Ginge in shot.

"What are we looking at here?" said Sam.

"We've crippled their fleet and now we're moving in for the kill," said The Ginge as he leaned in closer. "Did you see that? They've changed configuration. When they were going toe-to-toe with The Lore attackers, they spread wide, wider than the enemy craft. Now they're running at the mothership, they're funnelling into lines, one behind the other. It protects our numbers against the turrets."

He, Sam and the crew watched on, the cargo doors on the mothership opened again, releasing the last of their attackers. From above and below bright blue pulses of fire launched at the Fredahn fleet, this fire was matched from heavy friendly ships behind the attackers. "Brilliant, like clockwork."

"And you designed this simulation to accurately reflect how the enemy would behave in real battle?"

"That's right, as close as we can get it; they're pretty predictable, actually. Still scary, mind you, but at least it gives as a way to plan and beat them."

"But we haven't seen their attackers in action as yet, how did you mimic their movements and structures?"

"I've been working with a small team of the tactical military guys and gamers. We've been through days and days of footage from wars long ago. Back when the greys were at war with the humanoids and reptilians… before the council. We've used that as our template."

The hologramatic enemy mothership started taking damage, the last of its attackers fighting a losing battle.

"Surely things have moved on from then?"

"If I know anything about the Chardrekk, it's that they're not ones to change. They're pretty much OCD about it. They just think their ways are better."

"So far, that seems to have been the case."

"True," said The Ginge. "But they haven't had to deal with us yet."

He and the crew watched as the mothership smoked and faltered, its turrets all-but destroyed, its fleet protection gone. It listed, or manoeuvred, away from the incoming Baltwae forces. Either way, the results was the same. It took a couple more heavy pulse attacks before the hull was compromised, then a couple more before it was destroyed altogether.

"Boom!" said The Ginge, before letting out a whoo-hooo and doing a little celebratory jig. "Pwned."

"Pwned?"

"It's… a gamer thing."

*

Hensaro allowed himself a smile as those around him on the bridge of the Themmington buzzed with the noise and excitement of victory. Sure, the enemy was a mirage, a play of human trickery from the Universal Translator, but it represented a sure and triumphant step for the remnants of the Baltwae force.

They had developed from a rabble of lost souls into a formidable fighting force.

In front of him, holograms of the captains of the Corrax, Narran and Dorsano shared in the celebrations with their crews. He watched them enjoy the moment a while before speaking. "A profound victory."

"Great work everyone," said Cooper. "Brilliant."

Corrax captain, humanoid Detarion Anglarae made an unusual hand movement. "A significant indicator of our potential."

"I concur, we are ready," said Narran captain Kace Astoy.

"We shall not know the truth to that until we make our mark," said Hensaro Althot. "Celebrations are in order – not at what we just achieved but at what we have become – a fighting force, ready for war. Come, let us return to Baltwae. There are tales to be told, feasts to be consumed, drink to be shared and future conquests to plan. Besides, it is the Manchester Derby tonight and I must ensure City get the three points."

*

It had not been all work and no play for the boys and the crew in recent months. In fact, the isolation of the Baltwae outpost led many to feel compelled to engage in after-hours social activity as if their personal sanity depended on it. The upper level of the habitable decks, directly below the docking bays, evolved into an entertainment precinct. Species used various parts of the area to unwind in ways familiar to them.

It was the human space that dominated. Not only did humans outnumber any other species on Baltwae, it was soon discovered by all and sundry no other species delivered entertainment like those from Earth – sport, movies, music – nothing the other species could provide came close.

The human area was divided into a series of bars, clubs, theatres and restaurants offering the latest Earth had to offer. Not only was it a home away from home for most of the humans, many of the other beings soon became lured in by the latest EPL season, rock concert, blockbuster movie or TV series.

Hensaro was one of hundreds to adopt Earth's entertainment, finding himself drawn into sport – particularly several codes of football. He found himself a Buffalo Bills fan, a Manchester City fan and a St Kilda fan. Knuckles loved nothing more than sitting with him at the end of the week to watch a classic match of AFL.

Hensaro also made the observation that humans most certainly would have discovered galactic travel of their own accord several decades earlier if they hadn't focussed so much of their attention on keeping themselves amused. Knuckles thought he was probably right.

*

Sam and his crew followed The Ginge through the crowded entertainment precinct. The noise and hubbub of victory celebrations echoed through the entire outpost. The redheaded human could barely make a few steps forward before being stopped, thanked, high-fived or have his ear chewed off in conversation.

He reached the point where the entertainment decks opened up, the claustrophobic corridor giving way to a sea of revellers, bars, music, screens, holos and lights that seemed to go on forever. He breathed in the scent of celebration and nodded his head to the beat of the music. His mind contemplated how right the conditions were for ending his bad run with the ladies, before he buried such thoughts for fear of jinxing potential future conquests.

As he made his way forward he felt a force shove him from the left. He turned to see Edson Pelk, standing on a table next to him, grinning. Alongside the diminutive Kolprai his partners in crime laughed. "A fight with me would be your last," he slurred.

The Ginge and Edson stared at each other before the Kolprai joined his friends in laughter and offered the human a high five.

The Ginge reciprocated through the aroma of alien and liquor. He looked at Edson's sidekicks. "Don't let him drink too much; you know what he gets like."

"Anything for The Ginge," said Gwii Starn, before taking another shot. There was a short pause as the eyes rolled back in his head then he collapsed forward, landing face-first on the table he stood atop.

Edson and Fillopil Vargreet burst in hysterics, before Edson took Gwii's next shot and gave it to The Ginge. "To victory!"

"To victory!" said The Ginge before downing the drink then making an awkward gesture to camera team, still following him. "Soooo… gotta go, you know, say hi to the boys, chat with the captains, that sort of thing. We'll catch up."

"Why don't we come with you?" said Edson before slipping on some spilled liquid on the table and crashing next to Gwii.

"Mmmm… or you could stay here, for now, and we'll catch-up later... at some point."

The big frame of Gluff Hurn hulked up next to them. He held three pint glasses in his hand, serving as cocktail mugs, his smallest finger barely fitting through the handle. The drinks themselves were loudly coloured and accentuated with lurid straws, slices of fruit and cocktail umbrellas.

The Ginge smirked before Gluff glared at him in his customary silence.

"I'm sorry… I… figured you for more of a scotch kinda guy."

Gluff slurped the first of the beverages up through a straw over several graceless seconds. "Yummy!"

"It's probably the limoncello," said The Ginge. "OK, so, I'm gonna go over that way. You guys just… be here… enjoying yourselves."

He slipped away before they could incoherently respond, camera team in tow, moving on to his next round of dodging, weaving and hi-fiving. He turned to the camera as he walked. "Great guys the Kolprai, but they're not known for their restraint. It's probably not going to end well for them tonight."

He arced around a large group of attacker pilots, gathered in a circle and yelling loudly, some sort of dare or drinking game happening at the centre. Once clear of the obstacle he could see Cooper and the others. From his left Jasmine Jennings and Astrid Horn swooped in, yelling his name and wrapping their arms around him.

"We did it!" said Jasmine. "Whoo-hoo!"

"I saw, nice work."

"It was your nice work on the sim," said Astrid. "So real, Ginge. Unbelievable!"

"We're heading to the cocktail bar, want to join us?"

"Yeah, I'll be there. I've just got to catch-up with the boys for a bit."

The two kissed him on the cheek and went on their merry way. A wave of mixed emotions washed over him. What had just transpired must've looked great for his reputation on-camera, but the way it had transpired; the subtle signals they gave off, meant every essence of his being knew. He was stuck in a place no one ever wanted to be. He was in the friend-zone. Permanently parked, wheels clamped, keys removed from ignition and not going anywhere.

He took a second to compose himself and headed to the boys.

*

Cooper spotted Reynolds sitting alone beyond the revellers; he had that look about him, always on duty. Sure, there was plenty to do, there always was, but if he couldn't celebrate a milestone like this, when could he celebrate? It didn't do anything to ease Cooper's suspicion. He tried to ignore Oil Man's main guy, but couldn't shake Reynolds seemingly blinkless glare from his thoughts. He felt drawn to talk to him, compelled even. Now was the time, with training and morale at an all-time high and some Dutch courage under his belt, the captain was confident.

He arrived at Reynolds' side with a beer and a scotch, offering the latter to his military commander. Reynold's nodded his appreciation and nodded again inviting his captain to sit.

Reynolds took a sip, then studied Cooper with a raised eyebrow. "Glenmorangie, single malt?"

Cooper smiled.

"How did you know?"

"It's a captain's job to know."

They both returned the gaze to the celebrating crew.

"Why aren't you out there?" said Cooper, eventually.

"It's not my thing… too old."

Cooper nodded.

"No, I find you learn a lot more about people watching from afar," said Reynolds.

"And what have you learned?"

"That we are still a long way from being a disciplined fighting force."

Cooper nodded. "Maybe. It was a pretty significant day, though."

Reynolds maintained his focus on the masses at the bars. "You've done some good things here Captain. You have surprised me."

"Wow… was that a compliment?"

"It was an acknowledgement."

"In that case, acknowledgement acknowledged."

Cooper and Reynolds shared eye contact for a brief moment, something Cooper counted as a win, before standard programming silence resumed.

"You're not much of a talker, are you?" said Cooper.

Reynolds turned to his captain. "Sir, I appreciate what you're trying to do but, you and I… we're different, we want different things."

"I would hope we want the same thing."

"Things aren't that simple."

"Maybe… maybe not. Maybe everything's as simple as you make it."

Reynolds reached for his scotch again and the conversation lost momentum. After all his efforts, Cooper wasn't sure if he was seeing Reynold's military exterior or pressing on the influence of Oil Man underneath. He searched for the next angle in.

"So, do you feel the crew is ready?"

"Ready? No one can answer that until we face an actual enemy."

"Well," said Cooper, "we'll find out soon enough then.

"Indeed."

The moment was lost when Hensaro, flanked by his senior officers, approached the pair. "A fine effort Captain Simpson. You should be proud of what you have achieved with your crew in such a limited time."

Cooper looked at Reynolds briefly before responding. "Thank you, Captain Althot, we owe a big part of our preparation to you."

"And us to you. Today's simulation was a well-researched and executed piece of military strategy from all involved. Where is your associate The Ginge? I wish to thank him personally."

"He's heading this way, I think."

One of Hensaro's party presented the two captains and Reynolds with a tray stacked three deep with thin wooden vials all displayed at mesmerising angles – a work of art. Hensaro delicately took a vial from the top then looked at the humans. "Basharn?"

Reynolds looked dubiously at the offering but Cooper dived in, having long given up on asking what was in the concoctions he was presented with. If it was poisonous, Gandalf would let him know. Everything else was a pot-luck gamble between sickeningly sweet, ridiculously alcoholic or triggering some long-dormant part of the palate that should never be disturbed. He knew if he prepared his mind for the latter the resultant taste could only be a pleasant surprise. He carefully lifted a vial of his own and the three downed the liquid in unison.

Surprisingly, a pleasant, nectary taste filled Cooper's mouth. "That's nice!"

Reynolds raised his eyebrow approvingly after his, more cautious, sample.

Hensaro lowered his head in appreciation. "Basharn is sourced from the Satroff, a sacred creature on my world. This precious liquid is secreted from the reproductive glands."

"Ewww," started Cooper, before rescuing the noise into an, "OK".

To his side Reynolds reached for his scotch and swigged a healthy dose.

"My people deliver this tray of the rare liquid to your kind as a recognition of our thanks."

Cooper exchanged a look of devious intent with Reynolds. "I'm sure they'll really appreciate that." He turned to the crew around the bar. "Listen up people, drinks are on Captain Althot."

Within seconds they had been inundated with crew waiting to get their taste of the unique offering.

"You know, Hensaro," said Cooper. "It's traditional on my world that people make a toast in times like these."

"A crisp, bready product?"

"No, toast – a statement of words – like, something to give meaning to the drink we are about to have."

"A statement of words?" said Hensaro as he faced the gathered crew. "Very well human. How about – the enemy will regret the day they didn't finish their mission at Baltwae?"

There was a loud cry of approval from the crowd before a round of Basharn was consumed.

Cooper leaned into Hensaro's side. "That, Hensaro, was an outstanding first toast."

Hensaro smiled at Cooper then turned to the crew once again and upturned his vial for all to see. "May the power of the Satroff's crotch power us to victory!"

There was stumbling silence before the general consensus of 'ahh, what the hell' prevailed in another roar of appreciation. Amongst the celebrations Cooper turned to give Reynolds a nod, but only saw the back of him make for the nearest exit.

"What'd I miss?" said The Ginge as he finally made it to his destination.

"What," said Cooper, before releasing Reynolds from his thoughts. "Oh nothing, just a victory groin secretion toast."

"OK," said The Ginge over several confused seconds.

"Ahh, the great battle strategist The Ginge has arrived," said Hensaro. "A mighty effort for you and your team. Basharn?"

"Thanks Hensaro, you guys did amazing out there," said The Ginge. "Is that the groin secretion? I might pass… just this once."

"You sure? It's good for virility?"

"Maybe just one then… for team bonding and stuff."

Cooper rolled his eyes.

"Excellent," said Hensaro as one of his associates furnished The Ginge with a vial.

Knuckles moved in next to Cooper. "When are you going to tell him?"

"Tell him what?"

"What he's drinking."

"He already knows."

The Ginge threw back the beverage and smiled at Hensaro.

"There's something very wrong with you, my friend," said Knuckles.

"What?" defended The Ginge.

*

"Between everything that's happened and, whatever battle lies ahead. I think we're sitting on something brilliant here," said Pete as took a mouthful of crisps.

Jason nodded. "Agreed. Certainly from a general plot point of view, we've got good progression. Some of the footage we've captured his been unbelievable, it's just… I'm not sure that I'm 100% sold on the character stories."

"Like how?"

"Not sure, maybe I'm sensing some hesitation, like people are holding back."

"That's probably because you've got Sam and his team shoving their cameras into everyone's face every five minutes. The guy's a knob. I told you that was going to happen. We should just use the UT to get our footage."

"No, having the cameras physically on location is a great way to get reactions, working a treat too. No, I'm talking about something different."

"Does any of it really matter until we see what happens when we actually head to war? I mean, it's what happens then that will dictate what's important now."

"Perhaps. It might be worth reminding the others where they're at in the big scheme of things."

"Perhaps."

*

"Leave? Are you really sure this is the time," said Cooper.

"Absolutely," insisted Hensaro. "Three days to clear and focus the mind will do wonders for the crew."

"But we're at war, surely we can't have people jetting off all over the place, we've lost enough people as it is."

"Gentlemen," said Knuckles, as he wedged his way into the conversation. "Hensaro, you nearly ready? Kick-off beckons."

"One moment. We're not just letting them go anywhere, Cooper. We'll stick to the Carudian systems – light years from the trouble and over 70 worlds worthy of a soul's reflection time, whatever they desire."

"What's this about the Carudian systems?" said Knuckles

"Hensaro's talking about some R and R leave."

Hensaro smiled at Knuckles. "Which means you'll be able to see your family."

The Ginge moved into the conversation saving Knuckles from forcing a returned smile. "C'mon, we'll miss the start!"

"No shit!" said Knuckles, focussing on the matter at hand.

"Language!" said Pete as neared the group. "C'mon, kick-off time!"

"Ah, actually, no I wasn't swearing."

"Sure sounded like it to me, a grade B swearwords at that."

"On the contrary, The Ginge and I have been doing a little research. I think you'll find Noshytte is a settlement on the Rylockia, a small rocky planet in the Dartania system." Pete stared in dumbfounded silence.

The Ginge laughed. "True story. Interestingly, it's only three light years away from the planet Farquer, in the Gantray system."

"Seriously, what are you doing?" said Pete, coldly.

"Well, we didn't want to spoil your swearing thing," said Knuckles. "So, Lilly's helped us put together a list of names of things that sound like swear words but are, like, actual places."

"Farquedd," said The Ginge. He was looking at Pete and, after an extended pause, added, "Mining colony in the Klouru system is one of my favs,"

Pete hunched his shoulders. "Really?"

"Yep, we've got hundreds of them," said Knuckles, high-fiving The Ginge.

"We just cross-referenced all of your grade A, B and C swearwords with names of places in the galaxy," said The Ginge beaming. "Loophole, be-atch!"

Pete glared at the pair long and hard. "Morons."

"Enough," said Cooper. "Let's just move on and watch the derby. Should be a cracker."

"Noshytte," said Knuckles.

***

CHAPTER 25

"It's been a torrid time for the New Council forces since this war began," said faux Jamie, as he adopted his serious newsy pose for the imaginary Morning Show cameras.

"It sure has, Jamie," agreed the UT-generated version of Sarah, his partner in news crime. "Our man on the spot, Bryce Neagle, has all the latest in this special report."

Cooper watched the UT-fashioned reimaging of the Morning Show, present him with the latest on the galactic front.

Bryce Neagle looked down the barrel of the camera, the backdrop was infinitely familiar to Cooper. "I'm standing at the steps of the Galactic Council Chambers, Galactica – a place that has become a no-man's land in this truly epic war. Run by New Council, in territory that was formerly a Lore stronghold. Here, an uneasy peace remains.

"Council is not in session but the machinations of war are ever-present in this hallowed place. Voim 638, the being in charge, stands as a beacon of hope in a galaxy full of despair."

Cooper rolled his eyes, in part at the cheesy line, but mostly at how Voim 638's reputation had stayed intact after the events on Galactica.

"News of the loss of two more fleets has the New Council rattled; knowing new thinking and tactics will be required to turn the unending Lore tide of destruction.

"Whether those decisions are made here, in the war councils of our finest or on the battlefield, only time will tell. If they don't come, time will tell a different story.

"Bryce Neagle, Galactica."

*

"This better be good, Pete, you're eating into planet leave and, to be frank, denying the ladies of Acuban III the chance to meet a legend," said The Ginge in his civvies; Hawaiian shirt and board shorts.

"I hope you'll be handing out sunglasses," said Knuckles. "If you were any whiter you'd be translucent."

"Which is exactly why we're hitting the beach."

"We?"

"Am I late?" said Alan as he darted into the room, proudly wearing an identical outfit to The Ginge.

The Ginge looked him up and down, then groaned.

Knuckles burst in hysterics. "Awww, don't you make such a lovely couple. Honeymoon suite, is it?"

"Shut your head."

"Alright, enough," said Cooper as he leaned back on his chair and reclined his head on Rhiannon's shoulder. "Pete, what's this all about?"

Pete swallowed, unravelling any confidence that may have existed by those falsely assuming this would be a smooth, well organised meeting. "Yeah, um, OK, thanks Coops."

He looked at Jason, who nodded in encouragement. "Yeah, anyway, so as you know we've been putting together the dailies for the new movie and we've got the doco team aboard and everything."

He looked up expected a cutting comment from either The Ginge or Knuckles, but he was greeted with something far worse – silence. He looked back at his notes; papers stared vibrating in his hands. "So, anyway, we've got a few notes from your performances so far—"

"Performances?" said Knuckles.

"Sorry, I mean, how you're delivering your lines."

"What are you on about?"

Pete looked around the room and realised he had work to do to get the team onside, he mulled over how best to express his thoughts but could only muster. "Performance."

Jason cleared his throat in a somewhat too confident kind of way. "I think what Pete's trying to say is all of you are here on an emotional journey – that's the bit that audiences will relate to. So, it would be good, for example, to consider what your journey is, say, a few times a day. It's a great way to ground your delivery from an emotional standpoint."

"What the Farquer?"

"Knuckles!" said Pete, "That's the other main piece of housekeeping for today. It's not cool, man – there's no way the censors will pass your little loophole."

Knuckles looked at The Ginge. "Did he just say housekeeping?"

"I believe he did."

"Because it definitely sounded like he said housekeeping."

"Agreed."

"And I really didn't think Pete would be the sort of douche bag who said housekeeping."

"Well, I was giving him the benefit of the doubt."

"Guys, seriously!" said Pete through thinly disguised annoyance.

The was an elongated pause in proceedings, which eventually ended when The Ginge coughed and said the word 'douche' at the same time.

Jason stood over his chair. "Listen here, you little smartarse. Your friend here is putting in a lot of hard work trying to make this movie something brilliant – a piece of art. All he's asking for is a little help from you and if you weren't such an ingrate, you'd give it to him."

Knuckles also stood over his chair. "No he's not; he's trying to change how I do things. Things that make me, me!"

"Exactly," said The Ginge, also standing for effect. "And what's with the emotional journey shit?"

"Guys, chill," said Rhiannon.

Pete rose to his feet. "It's your personal story – the challenges you face within the overall story."

"And what exactly are my challenges?"

Pete paused and looked at Jason before responding. "You're dealing with a crisis of ego because your life revolves around getting the ladies but most of the people here think you're gay and even the women who know you're straight wouldn't touch you with a ten foot pole."

Knuckles burst into laughter and The Ginge felt himself go red. Alan went to put his arm around The Ginge but he slapped it away. Knuckles laughed again.

"Shut up Knuckles," said The Ginge. "Think it's funny, do you? C'mon then Pete, tell us Knuckles'."

"Enough!" said Cooper.

Knuckles looked at Pete. "Hang on Coops, I want to hear this."

Cooper found himself standing as well. "I don't think anything good's going to come from this conversation, can we just agree to limit the swearing for now and move on?"

"No way, I want to hear what my fancy emotional journey is."

"I really don't think that's a good idea right now," said Cooper.

"I think I have a right to know."

Everyone looked at Pete, whose eyes were fixed on Jason, for help.

"Well, for the sake of transparency," said Jason.

"The only transparency around here seems to be coming from Ginge's legs," said Knuckles.

Alan moved in to console the redhead but was greeted with a similar response as before.

"It's about dealing with loss," said Pete eventually to Knuckles. "You miss your kids, it's obvious."

The Ginge began a revenge laugh but it soon petered out as he realised the gravity of the situation.

Knuckles glared at Pete, already hurting but sensing there was more. "Go on."

"You stick to yourself and get stuck into work, but you're feeling left out there too because you really don't have the specialty skills of the rest of us. You're just struggling to find your place."

Attention in the room was now all focussed on Knuckles. "Excellent, well done, read me like a book," he said, clearly taken aback by the assessment. "Can I have a guess at your emotional journey... total ass-hat who would walk over his best friend if it gave him a better angle for his next shot."

He stepped back from his chair, met eyes with everyone in the room, turned and left.

Pete shouted after him in an apologetic tone but it was too late.

"Well done, Pete, you tosser," said The Ginge as he too got up to leave.

He'd almost made his way to the door, when he heard, 'wait for me' followed by heavy pounding footsteps. He hunched, waited for Alan to catch up, then left the room.

"You really need to sort that lot out, Coops," said Jason.

Cooper looked at him, then Pete. "Have you any idea how far down the priority list a movie is right now? If you do anything else to compromise members of the crew, I'll make sure it's the last thing you have to do with me, the Dorsano or the UFO4. Got it?"

Jason muttered to himself then stomped from the room. Pete apologised profusely before joining him, leaving Rhiannon and Cooper alone.

"Well, that was productive," said Rhiannon.

"How did they even think that was going to end well? I mean, Jason I can understand, subtlety's hardly his strong suit, but Pete? Wow."

"Don't even think about it," she said as she leaned in and kissed him.

Cooper enjoyed the moment then sighed hoping to let the weight of the world out with it.

"Do you know what I think you need?"

"What's that?"

"Two nights at the Intiina Spas Resort, Paramoca, Arranfell. Did you know the temperature only varies by two degrees this time of year?"

"Ahh, sounds amazing."

"So lose the uniform and put on some shorts, we are going to seriously let our hair down."

Cooper lent his forehead against hers. "You're amazing, Rhi, has anyone ever told you that?"

"All the time, as it happens."

"Really?"

"I kind of like him too."

Cooper laughed. "Really? Should I be jealous?"

Rhiannon stood, then offered Cooper a hand up. He accepted.

"Probably. He's smart, good-looking, caring, oh and did I mention he is captain of a starship?" she said.

Cooper embraced her. "Sounds like tough competition. I'd better make the most of this holiday then."

"I am open to being romanced in luxury."

"Consider it d—"

The door to the meeting room whooshed open and Dilania stepped through. "Pilot Bates said I would find you here."

"Everything OK?" asked Cooper.

"Yes, everything is fine."

"Then, how can I help you?"

"I hear, as fate would have it we are headed for the same destination."

"Sorry?"

"Intiina Spas Resort, need a lift?" said Dilania as she eyed Cooper. "I know a good pilot."

***

CHAPTER 26

Knuckles knocked on the door and breathed in the clean, chilled air while he waited. He turned his gaze to the mountains and enjoyed the freedom that came from being outside again. He heard the door open behind him and someone yell out, 'Daddy!' He recognised Balart's voice and a tear materialised in his eye.

"Hey little buddy," said Knuckles as he turned to realise his biggest son was anything but.

Balart was joined by Fjard and Michael, and Knuckles stared in wonder at how much they'd grown in the last few months. Scary. They rushed forward to embrace him; he prepared for impact, managed to stay upright then closed his eyes and breathed in their love.

When he opened his eyes again Wendy Wendy was standing in the doorway, leaning on the frame and taking in the scene. She seemed content – at ease. "When you boys have finished attacking your father you might think to invite him in."

They laughed and led him inside.

"Are you staying for dinner, Daddy?" said Balart.

"We're having querant," added Michael.

"We caught it ourselves," said Fjard as he posed in his tough guy's stance.

Wendy Wendy moved in and grabbed Balart around the waist, lifting him off the ground with one of her right arms. "Alright boys, give your dad some room to breathe, he's had a long trip."

"It's fine," said Knuckles, with a smile before turning his attention back to the boys. "Wait a minute, you caught a querant?"

"Down where the creeks meet," said Michael.

"But those things are nearly as big as me."

"And faster," added Balart.

Knuckles laughed again. "Thanks."

*

Alan came back from the bar with two large drinks, colourful and elaborate. "Get one of these into ya," he said.

The Ginge rubbed the sand from his hands and reached out for the beverage. "What the hell is this?"

"Orfon… no hang on, and Onafonti Orferto," said Alan with pride at his pronunciation. "Apparently they've got more kick than you'd think by looking at them."

"Is that because they look like girls drinks?"

"Do they?"

At that moment a couple of humanoid females strolled past in their beachwear. The Ginge nodded an acknowledgement and Alan ogled. The females looked at them, then their drinks before laughing and continuing on their way.

"Yes, Alan, yes they do."

Alan shrugged his shoulders then took a large swig. "They taste good though."

"Is that the point? We're not here to sip camp drinks; we're on a mission of love. Did you even pay any attention to the emotional journey stuff?"

He studied Alan who, it seemed, had completely ignored him, far more interested in sniffing the air. "What are you doing?"

"What? Nothing!"

"Smelling something – what were you smelling?"

"It's just; I got a waft of some sort of perfume from those two girls—"

"Dude, that's a little bit creepy."

"What? I'm a smell kind of guy."

"OK, that's even creepier. You haven't dated much, have you?"

"It's tricky in my world."

"Don't you MC sci-fi conventions?"

"And I have my own sci-fi podcast."

"Well, there must be heaps of fan girls thinking you can introduce them to the celebs."

"True," said Alan, "but those events usual end in a cosplay party."

"...and?"

"And, sure they can get pretty wild, but the guy to girl ratio is, like, four to one… more even. You've got to be on you're A-game. You'd struggle to find a more competitive market."

"If you'd asked me a few weeks ago I probably would've had some good advice for you."

"And now?"

"I'm on a losing streak," said The Ginge as he sipped his drink.

The two humanoids walked back past the other way. Again the two groups exchanged glances and the girls laughed.

"That might be about to end," whispered Alan as he stared at the ladies walking away.

"What are you doing? Don't jinx it!"

"What?" defended Alan.

"Don't call the end of the streak before it's happened, that's just bad luck. Quick, is there any wood? Touch wood!"

The Ginge looked at Alan still staring at the girls and sniffing in their scent once more. "Actually, bad idea. Please do not touch wood. Just, take it back."

"Alright, I take it back, Jeez."

It was only as the conversation tailed off and The Ginge, too, got a waft of the perfume that the idea came to him. The idea to change the war.

*

It's a God-like power having an access-all-areas pass to people's lives. Pete had never stopped to think of the ethical dilemmas wielding that power created. He just focussed on the next good shot, and telling the story.

The UF04 movie had received overwhelming praise and everyone aboard the Dorsano knew he was making another. Jason had even ensured everyone signed waivers, giving permission to use their image for the movie, documentary and any other related future spin-off.

Since his fallout out with the others, Pete had a new respect for the power he had at his editing fingertips.

Before, he'd happily use the UT to follow a crew member's every movement without question – trying to find the right moment to enhance their character arc or storyline, now, he felt guilty – creepy almost. He figured that feeling was doubled with most people on planet leave and the corridors of Baltwae's underground base all but silent.

He was nearly going to retire for the day when he noticed something – a look between Reynolds and Rodriquez during the rescue of the crew of the Gaujillo. Maybe there was a love story happening between senior crew members, definitely worth following. He saw it again. Then he noticed looks between the pair and Popov in engineering.

Now that was definitely intriguing.

He decided to investigate further and was diving into some footage when Jason entered the editing suite. "What are you working on?"

"Ahh, nothing," said Pete, suddenly defensive.

Jason studied the screen, seeing an image of Reynolds leaving the command deck. "Reynolds is a dead end, I reckon. He's too regimented; I think we're best to keep him as one of those two-dimensional antagonistic senior officer types. The bitter old school guy – we talked about this."

"I just thought I saw something."

"Waste of time, there's just nothing to add the depth required to make him a major part – no unusual ailment or family situation, no life changing incident – just years of tedious service to his country."

Pete looked at Jason, studying his words and formulating his response. Eventually, he nodded.

"The guy's a bore. Seriously."

"What about the Oil Man links?"

"It's such a thin thread, Pete. It's just going to take audiences out of the moment. It's a momentum killer."

Pete shrugged his shoulders then asked the UT to shutdown the editing dashboard. "I was about to call it quits, anyways."

"Good call. If you're looking for another subplot, did you see the footage of Knuckles and his crew save their survivors?"

"Not yet – still got a stack of stuff to go through."

"Well, when you get a chance, do it. Absolutely amazing. I've got a quick edit put together if you want to see."

"Nah... tomorrow maybe."

"Too easy," said Jason. "Hey, I'm not going to be long on this, you want to hit the bar later."

Pete stood up and rubbed his eyes. "Nah, I think I might have an early one, my brain's fried."

"Fair enough, catch ya tomorrow."

"Done," said Pete as he left the room.

Jason tweaked a couple of clips on his timeline, then went to the door to make sure there was no one outside. The coast was clear so he returned to his seat and dialled Reynolds on his communicator. It wasn't long before he answered.

"Hank, we've got a situation."

*

Cooper approached the bed with a plate of local fruits and nectars. He looked at Rhiannon, the morning Paramoca sun bathing her in a delicious orange hue. Her eyes opened and she gave him a complete smile, he sat next to her. "Morning, Rhi, breakfast?"

"Mmm, yes please," she said as she took a sip of nectar.

"Eat up, there's a big day ahead. The ferry heads to the atoll in less than an hour, we don't want to miss the—"

There was a knock at the door. Cooper left the food tray with Rhiannon and went to investigate. It was Dilania.

"Hello, Captain Simpson," she said, letting the words gush from her mouth like sighs of pleasure.

Behind him, Cooper heard Rhiannon sigh; the carefree vibe vanished in an instant.

"Dilania, what are you doing here?"

"We need to talk."

"Do you think it can wait until we get back to Baltwae?"

"I don' believe so, the subject matter is very.... sensitive."

"It's just; I'm spending some rare time with my partner and we—"

"Ah yes, the human need for coupling," said Dilania as she mocked a yawn.

"Yeah, it's just what we do, so if you'll kindly—"

"It concerns both you and Rhiannon."

"Oookkkayyy."

"There is a saying amongst my people: you'll find the best and the worst when you let people inside you."

Cooper looked at Rhiannon then back at Dilania. "Look, Dilania, you seem like a nice person and you're smart and talented. It's just, Rhi and I, we're quite, you know, content as a... twosome, we're not really looking for another lover—"

"I'm talking about the traitor in our midst."

"Traitor?"

"Or traitors, we – Hensaro and I – have not yet established that. But without doubt, the Baltwae attack was driven from the inside."

Cooper looked at both women in turn. "Who would do that?"

"The question has not left my head since the event, how did they know where we were – the odds are astronomical.

"We have conducted our research in secret – until proven otherwise we trust no one. But the task is proving too vast for Hensaro and I to complete. We need to bring in others we trust. Hensaro suggested approaching you; here."

Cooper looked at Rhiannon before returning his eyes to Dilania. "Thank you. What would you have us do?"

"We have completed research on 17 species; I am still to find anything. Here's the remaining list, if you take these, I'll continue with the rest."

Cooper looked at the list. "So, you've checked us then?"

"Of course, the fact you arrived after the conflict demanded it."

"And you didn't find anything?"

"No, but again, I am not surprised. Most species here have been galactically aware for many millennia, there are infinite connections to The Lore on many levels – personal, trade and political. Somewhere amongst the endless links is one that will still be active. It is there I am sure we will find an informant or unsafe connection. It is also not beyond the realms of possibility that the one who communicated our location did so without realising the consequences."

"So, what are we looking for?" said Rhiannon.

"I've been starting with the battle. Looking at each member of each crew and their behaviour in the lead up and their reaction as the battle commences. If we truly do have an informant among us, they will surely give themselves away in these moments."

Cooper and Rhiannon nodded.

"I must warn you this is difficult work, witnessing the same horror from so many perspectives can take a toll personally."

Cooper studied Rhiannon again before responding. "We're prepared for whatever it takes.

*

Knuckles tried to hold back the urge to burp but it was futile. The sheer volume of food he had consumed left his body no other option.

Wendy Wendy chortled and the boys were in fits of hysterics before trying to repeat their daddy's volumous act. Knuckles started laughing too, until he noticed one of Wendy Wendy's brothers, Vangillor, staring at him across the table.

It was not a nice stare.

Knuckles knew the stare was not for the act of burping but for the perceived lifetime of wrongs he had done to Vangillor's sister. He lowered his head in the knowledge eye contact with the plate would not cause it to want to fight him.

Wendy Wendy's mum, Blartoona, entered the room with a dish full of sweet smelling food. "What did I miss?"

"Daddy made mouth farts," said Balart.

"Well, he'll have more room for some Jurrow Tang then."

Knuckles looked at Blartoona, now standing over him. "I couldn't possibly eat anther thi—"

A large clump of the food was slapped onto his plate. He looked at it, then Blartoona before making the mistake of looking at Vangillor again – still staring him down. He rifled through the options of what to say next and the consequences, settling on, 'thank you' before eating a mouthful.

Despite its sweet smell there was an underlying flavour that Knuckles found far from pleasant. Had this been and ordinary dinner in an ordinary situation he may well have jettisoned the offending food into a napkin, but that wasn't an option today. He chewed and swallowed as quickly as he could without letting the discomfort show on his face. Once his mission was completed he looked up again and saw Vangillor staring right though him.

"So, Wendy Wendy tells me your people arrived late to battle."

"I... we..."

"They saved many lives," said Wendy Wendy. "They are heroes."

"A hero is someone who fights; a nurse is someone who helps the injured."

Knuckles knew that even if he had single-handedly fought off an enemy fleet and personally destroyed several large starships, leading to victory in the war, it would still not be enough for Vangillor. He continued to do what was expected of him and looked uncomfortable while offering limited eye contact.

Vangillor looked him up and down. "And what of your plans now?"

"We're ready to re-enter the war. Not sure if we'll join a larger fleet like the 79th or 83rd or see what we can do on our own."

"With four starships?" said Vangillor as he laughed at Knuckles. "So, let me get this straight, when you were 75 starships, you were annihilated, yet you think you offer something to the war effort as four?"

He stared at Knuckles, waiting for a reply; Knuckles again avoided eye contact, instead offering Fjard some help with his meal.

"You are as weak mentally as you are physically, human. While you're off playing war with your pathetic friends some of us will be taking the fight to the enemy."

Knuckles swallowed hard before uttering his next words. "And how has that worked out for you so far?"

Vangillor stood at the table, sending his chair flying backwards. It was such a sudden and violent move that all attention in the room fell to him. He towered above Knuckles and breathed heavily; highlighting the physical advantage he had over the human.

"Vangillor!" screamed Wendy Wendy.

Knuckles looked at his boys, they were scared. Aggression was such a fundamental part of male Gartorgian culture and Knuckles knew this was a defining moment for his limited hours of face-to-face fatherhood.

He closed his eyes for a second, thinking about the genetic footprint he had left his offspring. Because of it, his boys would spend their lifetime battling intimidation from the stronger full-blooded Gartorgs. On the other hand, they would have his brains; they may not be Knuckles' strongest suit in the human world, but more than enough to match wits with a Gartogian, if he dared. Today, he wanted to teach his boys, today he would dare.

"If you want to offer the enemy free target practice, be my guest."

"You have to be in the fight to be a target."

"On my planet, they say the definition of insanity is doing the same thing over and over and expecting different results."

Vangillor looked at his challenger, trying to understand where this unexpected bravado was coming from. Then he noticed Knuckles make a gesture towards his boys – a small notion, with his eyes, but enough to give his game away. He was posturing for their benefit. Gartorgians may not be smart but they are clever.

He eyed Knuckles again and smiled, knowing that beating the human in a verbal duel would be a far more hurtful blow right now.

"Really? And what exactly would you do?"

"That's... erm... confidential."

"Confidential? I see. You clearly understand warfare better than this lowly fighter pilot. Your 'arrive after the battle' strategy is still beyond my limited knowledge of battlespace tactics. Remind me of your rank again and role in such an intellectually superior plan."

"Vangillor, stop it!" said Wendy Wendy.

"Gunner," said Knuckles.

"I'm sorry, did you say gunner? Oh my, you humans really know how to bring fantastic philosophies to the war. Who would have though a gunner would have the responsibility for battle strategies?"

Knuckles looked at his boys, he had to do something. "I..."

"Vangillor!" cried Wendy Wendy as she mourned the loss of the chance at one normal family night.

"We have plans Vangillor, and we do things differently. And, for your information, I may be a gunner but I'm part of the team that runs that ship." He looked at his boys, "I'm positioned on the command deck."

"Yes, I'm sure you're the one they come running to when the big decisions need to be made."

"I don't have to prove myself to you, just know something big is coming."

"Is that so?"

"Yes," said Knuckles, doubt written in his eyes.

Vangillor studied the human with suspicion and smug superiority. "Well, I look forward to this, 'something big' to which you refer. I will be sure to monitor your movements to share in your glory, even if just to say that I, Vangillor, shared a meal with the man behind the victory in this war."

***

CHAPTER 27

Cooper and Rhiannon entered their quarters on Baltwae, exhausted from leave and the travel home. Cooper unzipped the top half of his uniform, rolled it down and let it hang from his waist before he slumped on the bed.

"Not so fast, my captain," said Rhiannon as she eased in next to him. "There still may be other duties to attend to."

"Yes, ma'am," said Cooper before smiling and putting his arms around her. "I'm actually surprised you're still talking to me after spending our planet leave with Ms Fatal Attraction."

"Funny thing is I actually enjoyed it – using the UT like that. Watching the same event from so many perspectives, following people, their connections... you can delve so deep. That's a pretty powerful thing."

"Pity we didn't find anything."

"I'm kind of glad we didn't."

Cooper though on her words for a moment. "True."

He rolled Rhiannon on to her back then knelt on top of her. "But seriously, thanks for, you know, not letting her get to you. She can be pretty... intense."

Rhiannon ran her hands down his chest. "You're just lucky I have faith in my captain."

"Interesting. Did your captain happen to mention your body does things in that uniform that make him have very unprofessional thoughts."

"Is that right? Well, perhaps we could do something about that. Something unprofessional."

Rhiannon took Cooper's hand and placed it on the zipper that sealed her uniform. He moaned his approval and guided his hands down the line of the zip.

Cooper's communicator burst into life. A light pulsed out and a pleasant and chirpy chime sang through the air. Cooper looked at it, swore, then returned to the task at hand.

"You going to get that?"

"No," said Cooper before kissing her body. "No I'm not."

*

"He's not answering," said The Ginge as he put down his communicator and picked up his beer.

"What do you mean, he's not answering?" said Pete.

"It's pretty straight forward – I pinged him, he didn't respond."

"But this is Cooper – he always, always answers. He might still be in transit."

"Nup. Got back a good hour ago."

The boys sat in silence for a moment. They sat in an observation room overlooking the docking bay in full swing. Every minute seemed to bring a new craft returning from planet leave.

"Well, I'm not letting a good beer go to waste," said Knuckles as he finished his beverage and picked up the spare on the table. "So, Ginge, how did your lover's holiday go?"

The Ginge took a large drink. "Two words: Dismal failure."

Pete and Knuckles laughed and clinked glasses.

"Ouch!" said Pete.

"The streak continues," said Knuckles though a thinly disguised grin.

"Streak!" said the Ginge. "Don't call it a streak – it becomes a thing when you call it a streak."

"How else would he refer to a continued sequence of weeks without getting any?" said Pete.

"Just... don't even refer to it."

Knuckles looked at Pete. "Does he seem a bit touchy to you?"

"Definitely touchy."

"I am not. Besides, it's not me, it's Alan. He's the galaxy's worst wingman."

Knuckles laughed. "It's Alan's fault! That is priceless."

"I'm serious. He followed me everywhere and he was loud and confident yet was way too desperate. Seriously, his swag to ability ratio was at least—"

"His what?"

"Swag to ability ratio – SAR – the difference between how good you act like you are compared to how good you actually are."

"Is that a thing?" said Pete.

"I'm making it a thing – and lumpy Alan gets a SAR of 7-1. He may as well have held a sign over his head saying 'desperate tool seeks live being for three seconds pleasure'. Meanwhile, I couldn't get a word in edgeways. I'm telling you he's, like, my sexual kryptonite."

"Dude, you really are in a rut," said Knuckles.

"Don't say rut, that's worse than streak."

"Hang on, hang on," said Pete. "So, if he's your sexual kryptonite, does that mean you think you're a sexual Superman?"

The Ginge's face went through a range of emotions as he contemplated the statement.

"Oh my God, he does."

"I do not!"

*

Walter Johnson sat in his living room, sipping a cup of tea and admiring a 3D display that beamed out of his communicator and floated above his dining table.

The scene was of the developing moon base. In a few weeks it would be complete and a small team of humans would move in to oversee the manufacture of the sister ships to the Dorsano – the official ones at least.

The production process was more wonder than genius to Walter. The surface of the moon was deconstructed and manipulated at an atomic level to become the building blocks that made the base. Every moment he wasn't working on the project – or the secret second project – he would watch the building progress. No human hand was involved; the construction was the work of a program, instigated by the UT, to turn the designs of the base into a functioning reality.

He took control of the display and zoomed in around the base, flying by as walls built up and over the lunar surface. He laughed at the power he wielded, at what he was part of; at a lifetime of dreams being surpassed in their twilight. He laughed until he cried.

He wished his wife was still around to share it with.

His communicator bweeped into life. He took a handkerchief from his pocket and dabbed his eyes before answering. It was Cooper. "Hello!"

"Mr Johnson, how are y..." Even through the hologram reproduction, Cooper could make out the redness in his former science teacher's eyes. "Are you OK?"

Walter laughed before dabbing his eyes again. "OK? I haven't felt this alive in... years."

Cooper returned the smile. "Well, I was just checking in to make sure I hadn't dumped too much work on you."

"It is no work at all, literally. Most of the time I just sit here and watch it happen." Walter gestured to the hologram floating behind him. "It's wonderful, Cooper."

Cooper smiled. "...and our other... situation?" He was confident Walter knew enough about the UT, and the people they were dealing with, not to let any details of their secret project slip, but he started in obvious conspiracy speak regardless.

"Equally as pleasing, I understand."

Cooper smiled; it was nice to know he was in trusted company. "And Anne? Working out for us, you think?"

"Very well, you've chosen well there. She's a smart operator, that one. Taught me a thing or two about dealing with some very important people."

"And how goes that part of the process?"

"My least favourite, I must confess."

Cooper nodded and Walter didn't need much encouragement to elaborate. "Well, I would have thought the project rather straight forward. With the ships, we are replicating the Dorsano's design and with the moon base, it's a design pulled straight from the UT commons. We select what we want; where we want it built and press the button. But, it seems every government and interest group powerful enough to have a say is, well, having a say."

"There's a surprise. They're so desperate to, I don't know, have an impact or something. It's pathetic. I am not sure what to say."

"It's not your fault, son. Clearly, we're blessed here on Earth. Our leaders are so clever they even know how to improve the design of alien spacecraft."

Cooped laughed. "Yeah, they are uniquely talented."

They each stood in their respective spaces on opposite sides of the galaxy, eyeing the hologram of the other and smiling. It was a smile of shared respect and trust. A smile neither of them had used enough recently.

"Listen, Walter, I'm... gonna be difficult to get a hold of for a little while. Impossible actually. Are you alright to keep doing what you're doing?"

"Let me see, helping build mankind's first moon base and first fleet of light speed worthy space craft... or retirement. I wouldn't have it any other way."

"Brilliant."

"Although, I suspect you'll owe my golf handicap an apology when this is all said and done."

"Tell it I am deeply, deeply sorry."

*

"Essentially they return to the gathering point then disappear to their various home worlds," said The Ginge to the fleet leaders back on Baltwae.

"And you discovered all this by following their... smell?" said Hensaro.

"Yeah, pretty much. Each ship in the void emits its own unique combination of exhaust – a combination of their drive technology and make-up. The ship's exterior also contributes, small amounts of matter tend to leach with travel. It's a very subtle trace and, while many ships are similar, the precise combination made by the super fleet may as well be a gigantic neon sign."

"A what?"

"Just... doesn't matter. What matters is regardless of how well they cloak themselves from detection using the UT; we can spot them when they gather ready to attack."

"You can see the invisible."

"Yep," said The Ginge with a beaming smile. "Like, farts on the breeze."

"Really?" said Cooper.

"But it gets better," continued The Ginge, ignoring his captain. He directed everyone's attention to the hologram presentation in the middle of the room. "Many of the ships are one-off builds – unique designs with unique emissions. This means we can track their movements, beyond when the fleet disperses."

Seventeen starships were displayed.

"We know exactly where these ships are based between missions, including this guy."

The Ginge paused for a moment while he engaged the UT through thought. All but one of the ships being displayed disappeared from view, leaving the largest craft. "Ladies and gentlemen, I give you the Cremmerson, hailing from Traigh."

"A Chardrekk ship?" said Hensaro.

"The Chardrekk ship. This baby is the mothership of the entire operation."

"Outstanding."

"From what we've found so far, the captain, Isla Enchant, is coordinating the attacks. If we take down the Cremmerson we will put a massive hole in their operations."

"Well done, The Ginge, this is precisely what we've been looking for," said Hensaro. "Members of the Baltwae fleet, our mission is clear."

"We can't attack the lead Chardrekk ship on their origin world. That is suicide," said Erranj Meld.

"No," said The Ginge. "But it docks at an outpost in the far reaches of the Traigh system, where the crew disembark to head to the inner system. There's only a token force based there along with a small resident population – totally vulnerable.

They all studied the data that accompanied the visual of the outpost; a modest amount of craft performing a slow dance around a space station, itself modest again the moon it orbited, all of which was dwarfed by the planet it circled.

Hensaro studied the image on the screen in contemplation.

"Leader, you are not considering this course of action?" said Kace Astoy.

"The enemy is strong and confident beyond words – they attack without concern. To strike such a significant blow at the enemy's heart will have meaning beyond the damage caused. It will make them think, it will slow them down."

"But, leader, we know nothing of the Cremmerson and its capabilities. Attack is foolish."

Hensaro glared at Astoy, who did not maintain eye contact. "I could not agree with the commander any less. Our friends from Earth have delivered us very significant information – the sort that turns wars. To have this data and not attack, that is foolish."

Astoy nodded in obedience. "As you wish."

Hensaro looked at the other leaders in the room. It was a gesture that encouraged others to voice their opinions, but prepare for the fact the plan was in motion. When he was greeted with silence he continued.

"How often have the attacks been occurring?"

One of Hensaro's subordinates consulted with the UT. "With this fleet, every 4-5 days, leader."

"And our forces? Are we ready for such a conquest?"

"We do not know their weapons capability, military numbers, fighter numbers, or craft designs, nor make up of those craft – we know nothing. Yours is a difficult question to answer," said the subordinate.

"Then, it remains something we do not know. We can assume they will have superior firepower," Hensaro eyed Astoy, drawing the doubt out in his expression. "But we have surprise on our side."

Hensaro broadened his presentation, to address the other leaders. "History is filled with the unlikely – the few – overcoming greater odds and numbers than we will face, and winning."

He lifted his chest and spoke with an authority that made Cooper feel he could fight to the end of the galaxy. "People of the Baltwae Fleet, leaders of worlds. We have been blessed. The great ether has, in its wisdom, bestowed pain upon us, yet given us advantage. We are a forgotten force, yet we are nimble, well drilled and..."

Hensaro approached The Ginge, acknowledging him with a nod. "...we have critical intelligence. In this war, in the way it has been shaped by our enemy, we have been handed the perfect combination for success."

Hensaro eyed the other leaders. "Do I have your weapons alongside my weapons, your people alongside my people, your soul alongside mine?"

There was a large explosion of noise – the collective agreement from the leaders of worlds. So loud and inspiring was the noise, Cooper could barely hear the sound of his own voice amongst the din.

Hensaro smiled, then laughed. "I thank you all, your heroism is no surprise to me. From this moment complete secrecy is essential. All communications beyond Baltwae are no longer acceptable and will be blocked – inform your people. Today we return to the shadows that should have veiled us since the beginning. We shall continue routine with our bodies but prepare in our minds, knowing any moment – soon – we will detect the super fleet gathering once more. That will be the chiming of our destiny. We will prepare and launch all craft towards the outer system around Traigh, target: the moon of Hazsis, Zoesam. We will stay in the shadows until The Cremmerson arrives. We will make ourselves known and fight for those lost at Baltwae, for our worlds and for the future of the galaxy. We will not rest until the enemy is defeated or we are destroyed!"

There was a deafening roar – a combination of united pride and patriotic certainty mixed with fearless uncertainty. They were together – one way or another – until the end.

Hensaro breathed in the atmosphere like it had a mainline connection to his soul. "They will remember our names, you know, they will remember this moment."

*

Hensaro's words echoed through Cooper's head for the rest of his time on Baltwae. Time wasn't time for that period, it was no time. While each battle sim they conducted made them feel a little closer in style and understanding to each other, and each additional piece of recon gave them additional insight, none of it truly mattered. They were on a collision course with a life changing event and all the in between moments were nothing more than deciding which black suit to wear to a funeral.

***

CHAPTER 28

Cooper looked around his quarters. The neat space, while comfortable, never really felt like a home. Captain's quarters – the sheer weight of the name alone meant he couldn't truly relax. He was always on call. Outside the entry doors, his crew busied themselves in preparation for battle.

He looked at Rhiannon lying in the bed, so peaceful. Then he turned his gaze out the viewshield, where he could see the lights of the Baltwae base in the distance. Closer to the Dorsano, the Narran lurked like a goliath, its gleaming exterior belying the recent trauma it had seen.

He paced back and forth, feeling slightly uncomfortable at how he was going to approach what he was about to do. "Gandalf, are you there?"

"I am everywhere."

"I know, I know," said Cooper, letting the comment drift off into cold dark skies above Baltwae. He started pacing again.

"Do you seek something of me?" said the UT.

"What? Umm, yeah. I want to do a... a thing."

"Thing?"

"I can't believe I'm doing this. I want to do a captain's log, a journal of my thoughts and stuff."

"And how can I assist you?"

"I'm not sure actually, I just... just... if you could archive them for me."

"All of your conversations are archived in my systems."

"I know... sorry, feeling a little self-conscious about the whole thing. I'm just blathering, ignore me."

Cooped scanned the room for the best place to deliver his first message. The low ledge at the base of the viewshield – it was his favourite place to sit and the space beyond would make an interesting backdrop. He sat and focussed his gaze on the sleeping Rhiannon as if imagining he was talking to her would make him less uncomfortable. He took a heavy breath.

"Captain's log... umm, number one."

He shook his head at the stupidity he felt. "So we're preparing the Dorsano for battle. We've been training on the rogue planet of Baltwae for several weeks now, with a division of Fredahn, Kindria and various others. Well, we were a division before The Lore wiped most of us out before our training began.

"In truth, we're not even a human handful of starships and just an underprepared group of survivors really. The training's been good; we've come a long way in a short space of time. The battle sims have been increasing in difficulty and the crew have really stepped up – coordination, communication, chain of command – everything. We're not great, but we're a world away from the early sims back on Earth.

"They're good people, talented people and I'm pretty proud of what we've created out of nothing."

He smiled at Rhiannon's sleeping body.

"In a couple of hours the Dorsano, as well as the Corrax, Themmington and Narran are heading behind enemy lines. We've devised a technique to detect movements from the enemy's super fleet. Right now, that fleet is gathering to destroy another one of ours. We have no idea what or where, but we do know The Cremmerson, the largest craft in the Chardrekk fleet, is amongst them. We're pretty sure that's the ship calling the shots... or at least the people on it are.

"We also know once they've taken out their target, whatever that is, the Cremmerson will be heading back to the Traigh system – home of the Chardrekk.

"So, we're heading behind enemy lines and going to wait for her return before we attack."

Cooper laughed at the hopelessness of it all. "It's probably not the most sophisticated plan, but it will send a message that we're not going to be messed with.

"Are we ready? Maybe. You know what, I actually have no idea. For all our sim work we've never actually fought anyone. The Ginge tells me the AI in the sims is first rate but nothing beats experience and, well, we have none.

"It's like picking up a new game, doing campaign mode and thinking you're a star, then jumping online and playing against actual humans. They think differently, react differently – thought-patterns, reaction times and just the general randomness of a live being. You can't train for that.

"The only thing that gives me hope is, from what we've seen of the mega fleet, they haven't had any experience either. So far it's all been one way traffic, heavy fire from the cruisers to destroy our starships without even deploying their attack craft."

Cooper noticed Rhiannon's eye open, she watched him in silence.

"I'm gabbing on," he continued. "I guess what I'm saying is, I don't know whether I'll be here in a few hours from now, I don't know if any of us will.

"I hope to be posting another one of those soon, talking through our victory. If not... to my family – I love you... and I'm sorry. To my crew – I thank you all. To the people of Earth, you should be proud of what we have done here. Don't let it stop with us; you will soon have a new fleet, let it be the start of a new era on Earth.

"Captain Simpson. Out."

He locked eyes with Rhiannon, each expressing the apprehension in the situation and pride they felt in each other.

"So, Gandalf, if the worse does happen, can make sure that message gets to my family? Oh, and Walter?"

"It will be done."

Cooper returned his gaze to Rhiannon, who tapped the empty bed space in front of her playfully, summoning him over. He smiled and moved onto the bed next to her.

She studied him; even the subtlest expressions at the corners of his eyes and mouth told her he was wearing the weight of several worlds on his shoulders. She ran her hands down his face. "It's going to work out, Coops, I know it will."

His mind raced to find the right way to handle his response but Rhiannon was having none of it. She put her index finger under his chin and gently pulled his face to hers. They kissed.

"I love you," she said as they kissed some more. "Gandalf, can we have some privacy?"

"As you wish."

The view to Baltwae and beyond disappeared, the window becoming solid wall. The captain's quarters became Cooper and Rhiannon's private space. They shared each other in ways more intimate and powerful than they'd known. The sort of passion that can only be achieved on the edge of the abyss, the first time... or the last.

*

"About time," said The Ginge when his captain finally emerged from his quarters. "I've been waiting here for ages."

"Like you had anything better to do."

The Ginge's mind raced as he searched for a sharp retort but fell tragically short. He watched as Cooper and Rhiannon marched past him, headed for the bridge, then he shuffled around and caught up to Cooper's side.

"So I think we're all set. We've transferred some of our fighters to the other ships to even out the attack numbers, but I think we're looking pretty good."

"Nice work Ginge."

The Ginge jumped up and down as he walked. "Ahh... can you feel it? Game day, my friends, game day."

Cooper studied his best friend – totally in the moment, all expectation and zero consequences. Cooper smiled and slapped him on the back. "Let's do this."

"Captain on deck," said Jasmine Jennings as the trio approached the bridge.

The Ginge sniggered just has hard as he had the first time he heard the New Zealander utter deck in the dodgy way she did. Cooper resisted the temptation to roll his eyes at his mate; instead he looked at Jasmine and gave her an appreciative smile.

Cooper stopped just short of the bridge door and looked at Rhiannon and The Ginge in turn. He spoke no word aloud but his eyes said 'we're in this together, whatever the outcome'.

"Game day," he said eventually before turning to the door.

It whooshed open as Pete and Knuckles whisked through at speed. They skidded to a halt as they saw the trio on the other side.

"About time!" said Pete.

"Seriously, we were about to report you to missing persons," added Knuckles. "What the hell have you been up to?"

Rhiannon blushed before regaining composure but Pete saw her guard drop long enough to join the dots. He looked at Cooper. "You sly dog."

"Focus lads, this is it," said Cooper.

"Wait a minute, have you two been..." said Knuckles, finishing his question with a raising of his eyebrow.

"Looks like it's not only the captain's who's been on deck this afternoon," said The Ginge.

This time Rhiannon seriously blushed.

"Really," said Cooper. "You're going to do this now?"

"I could say the same thing about you two."

Cooper looked around at the crew watching on, ensuring his words stayed out of earshot. "Alright, enough. You guys ready to kick some butt or what?"

They all stood in silence, like told-off school children.

Cooper let them hang uncomfortably for a moment. "Good. Alright then, perhaps for once, we could act like professionals. I mean, we're only leading a starship into battle deep behind enemy lines. Perhaps we can put the dick jokes and innuendo on hold for the afternoon, walk into that room and, I don't know, at least pretend we're capable of the task."

Again they stood like scolded schoolboys, until The Ginge sniggered.

"Really?" said Cooper.

"I'm sorry, I'm sorry. It's just, there's something really funny about the word innuendo." The Ginge breathed out hard and slow, trying his best not to let the giggles take over.

"Have you finished?"

The Ginge, still unable to talk, nodded his apologetic approval.

Cooper shook his head and turned to Pete and Knuckles. "And you too."

Again he was greeted with more watery eye stares.

"Wow," said Cooper as he focussed on Pete. "Captain's log... two. I am surrounded by morons. Ends."

Eventually the smiles subsided. "OK. Any danger we can actually do this now?"

"Yes, sir!"

He looked at his friends, shook his head once more then turned his attention to the bridge door. "Game on."

*

The planet Mersai was almost entirely unremarkable and completely ignored by the sentient beings of the galaxy. Its bulky mass and toxic gases made it completely devoid of life and it sat in an off-the-grid corner of space. Its dozens of moons danced around it in a finely-tuned choreography crafted by the eons to an audience of zero.

On most days.

On this day, the green, blue and purple world and its several satellites were witness to more action than in it had seen in the many millions of years before it, combined. A fleet of New Council craft gathered, awaiting their next deployment. The fleet was smaller than usual – only 27 craft. The leaner size marked a shift in strategy for New Council.

Smaller, but not too small to go ignored.

The fabric of space burst open, shooting a large fleet of The Lore ships. One by one, they came through as if skidding silently to a halt in silent synchronised formation. As the numbers continued, the fleet formed a spherical structure, at its centre, The Cremmerson.

The leviathan bulked into position, dwarfing the star cruisers in its company. Its sleek, dark grey exterior shimmered as it reflected a light show from the other craft and the local dwarf star.

At the helm of the Cremmerson stood its captain, Isla Enchant. She cast an imposing figure of height, rank and experience – assured and determined. She stared at her prey through the viewshield and smiled.

"Connect me with their commander," she ordered.

"Issuing communications protocols," said one of her underlings. "Senior commander De Fa Linguey being summonsed."

Soon, the hologram of a humanoid appeared in front of Enchant and her crew. She looked the being's likeness up and down, her disgust at the humanoid's tiny eyes and primitive hair visible. "I have a question."

De Fa looked back at the grey, her eyes telling all who watched she knew her life and fleet had mere moments left. "Will it affect the outcome of this encounter?"

Enchant glared back. "No, but is it not the task of a leader to understand their enemy, no matter how unsophisticated?"

She touched her elongated index finger on her chin as if in mock thought. "Do you value the lessons of history, humanoid?"

De Fa stood in silence for a moment as the crew around her scrambled to ready for battle. She was taken aback, trying to read between the lines for a hidden intent. "I... yes, I do."

"Excellent. Then, as you would be well aware, my species was the first to evolve and discover space travel, we were the first to populate other worlds. History tells us it was millions of years before your people reached the same discovery. And what is the first thing you did when you became strong enough? You started a war. A war that lasted centuries and cost countless lives."

Enchant leaned forward as if wishing to reach into De Fa's soul. "We should have finished your kind then and there. My ancestors, in their wisdom, and perhaps in hindsight naivety, chose to offer peace and a new galactic order. We gave you access to our technologies, showed you a deeper level of understanding of the universe; we gave you everything. I consider it a sign of our great character as a species, even if the decision ultimately led us to this moment.

"Now we know, with complete certainty, you must be destroyed."

De Fa locked eyes with Enchant through their hologramatic connection. She knew her fate was sealed, yet she searched for the right words to counter. "I believe the history that you speak of so fondly will show the New Council only sought to deliver peace and equality to the galaxy."

Enchant laughed. "That is a very interesting perspective. I would love to know how you arrived at such a confused interpretation."

"The last council—"

"But I don't have the time, nor inclination, to listen to you further. Before I destroy you, I shall give you one more lesson about history – an important lesson. It is written by the victors."

De Fa watched the face of the grey turn dark, her large eyes narrowed. "Isla Enchant, please!"

"Commander Mersion, Destroy them," Enchant ordered.

"Attack!" screamed De Fa as comms were terminated.

Isla Enchant stood at the bridge of the Cremmerson and watched as pulse fire approached from the enemy. She smiled as her ship and the scores of cruisers that surrounded it returned serve on a scale that dwarfed the enemy's attack.

It was over in minutes. The victory for The Lore overwhelming, as had been the many before.

Enchant smiled, she would soon return to Traigh, delivering her people one step closer to their goal, her status amongst her race assured.

Commander Elrus Mersion approached. "As you commanded, the enemy are destroyed."

"Excellent."

"High commander, why do you engage them in dialog? It gives them precious seconds to prepare."

"We have far too much firepower to be concerned with such things. I am merely allowing them a chance to fear their demise and reflect on their mistakes. I give purpose to their death."

She eyed her understudy. "You have a lot to learn yet Commander Mersion. Now, let us return home."

*

Captain Cooper Simpson entered the bridge of the Dorsano. In his wake were Wesley 'The Ginge' Ellis, Pete Bates, Bradley 'Knuckles' Thomas and Rhiannon Richardson. The others' fanned out to their positions on deck, while Cooper headed to the helm. He looked across the bridge – all eyes upon him.

"Gandalf, are you there?"

"I am everywhere."

"Sure. Hey listen, I just want a say a few things before we depart, can you beam this out to the rest of the crew."

"It is done."

Cooper nodded to air, collected his thoughts and looked at the faces of his senior crew in turn. "I'm not sure what's going to happen today. I know it's going to be something big."

He laughed to himself. "I also know that the crew that heads out today to represent Earth is a world away from the one we dragged together a few short months ago."

There were pockets of laughter among the crew. "I think we were more a danger to ourselves than The Lore. But, when you think about where we've come from, what we've done... and what we're about to do... it's nothing short of a miracle.

"I want to thank you all personally. We're the first step in changing our world for the future... and for the better. That's a legacy that will last many, many lifetimes. You all deserve to be part of that.

"Today we're going to hit them on their home turf. They're not going to see it coming. If all our recon is correct, we should, at the very least, match their numbers. And with the training we've been doing..."

Cooper let the sentence drift into the void while he studied the faces in front of him. "I think we can take them today. No, I know we can. I'm convinced the Dorsano will be heading back to Earth with victory under its belt."

The crew hollered, whistled and applauded. Cooper pumped his fist in front of his face. He knew they were all as unsure as him but between his speech and their response, it didn't matter. It was as if they all knew something bigger was at play here than just victory, or their lives. This was a moment in history – a statement. It was not about personal survival logic, it was a time for sentiment on the greater good – a time to rally as a part of the future of humanity.

"Mankind will remember this day forever. Let's make this day everything it deserves to be. Let's kick some arse!"

Another wave of celebrating filled the bridge, the wall of noise even bigger than before. At the heart of the space, Cooper stood over his chair, taking it all in, smiling.

"Pete," said Cooper over the noise. "You got this baby ready to rock?"

The crowd hushed as if witnessing this moment as an observer for posterity.

"Affirmative, Captain. The Dorsano is good to go, sir."

A few clapped while someone whistled a sign of appreciation.

"Excellent," said Cooper. Once more he felt the focus of all in the room as he lifted his eyes to what lay in the space beyond the viewshield. "Gandalf, can you get Hensaro on the communicator?"

"Yes."

Seconds later, the Fredahn's likeness appeared on the bridge. He cast his eyes around the space, eventually settling on Cooper. "Congratulations, you look prepared my human friends. It is good to see my old ship in capable hands."

"Thanks Hensaro. There'll always be a spot for you on this bridge."

"Then victory is all the more worthy of fighting for today, good friend."

They shared a smile. It was an act of solidarity to hearten the crew around them, yet, lurking inside the look were discreet tells of doubt, subtle enough that only the captains perceived their meaning. It joined them closer again and soon the doubt was replaced by knowing they belonged in each other's company – leading this force.

"Well Captain Hensaro, we're at your side and ready to depart."

"Excellent. Chief Leader Kace Astoy?"

"GX-124 Narran at your side."

"Commander Detarion Anglarae?"

"The Corrax ready and willing to serve with you, sir."

"Acknowledged. Let us not waste any more time, let's change a war."

"Yes, sir," said Cooper as he saluted the Fredahn.

Hensaro's likeness, and that of the other captains, disappeared from view and Cooper looked around the bridge. This was it.

He'd reached the moment – the anticipation tipping point. His nerves had grown from when this mission was devised until this moment. Now they swelled to fill his stomach with a bottomless pit of dread. Then Pete relayed a message from the Narran and Cooper nodded his approval for the Dorsano to depart. He took his seat and in an instant, the nerves left him.

His sole focus became the battlefield and victory.

***

PART 4:
BATTLE

CHAPTER 29

Cooper watched a kaleidoscopic tunnel of tubular light barrell into the horizon as the Dorsano ripped through the fabric of the galaxy.

Just as quickly as the show started, it came to an end as the Dorsano, Narran, Corrax and Themmington eased their way back into a more easily perceived reality.

They were headed toward the fiery gas giant Zoesam, farther in the distance a small red star appeared. It was the first natural light they had seen in a long time but no one had the inclination to linger on the moment.

"Can you see the outpost, Pete?"

"Not yet. Near the bigger of the moons apparently – Hazsis; heading there now."

"Excellent. Remember – silent running – no comms 'til we engage."

Pete nodded as he manipulated the controller.

"And stay in tight on Hazsis. No need to show ourselves before we have to."

Alongside the Dorsano, the other craft eased forward in silence. The tension so heavy that the silence was even observed on the bridge – as if somehow talking would alert the enemy. The moment stretched out across the horizon like light speed, eventually broken by Knuckles. "I think I saw something."

There was a distortion in space by the large yellow moon and soon the Cremmerson lumbered into view. Though still far away, there was no mistaking its imposing physical endowments.

"Oh my god, it's huge!" squealed Pete.

Hank Reynolds was in dialog with someone on the other end of his communicator. "Fighter crew loaded and ready, sir."

"Thanks, Reynolds. Knuckles, your team ready to go?"

"Affirmative – all turrets manned... or womanned... or whatever."

"Thanks, Knuckles. Pete, how are we looking?"

"So far so good. Don't think they've seen us yet. If we can stay in the moon's cover, we should be able to get pretty close at this rate."

*

Isla Enchant smiled as her ship eased into the comfortable space of her home system, victorious once more. Nothing beat the feeling of returning to Traigh a hero – soon she would saviour the accolades of success.

There was a commotion by one of the tactician's displays; soon the distraction had escalated to the point where she needed to be interrupted.

"We're getting an usual signal, Captain," said the tactician.

Enchant sighed. "What is it?"

Second in command Elrus Mersion, looked down at the tactician, throwing all the responsibility on the underling to deliver the news, or the incorrect read-out.

The minion looked at Mersion, then Enchant as a wave of doubt and dread washed over him. He gulped. "At first I thought it was nothing, but I think there may be a problem."

"Problem?" said Enchant, low on patience.

"Unusual movement high commander, beyond Hazsis. Stand by."

The silence was choke–inducing as the minion urged one of his colleagues, currently buried in the UT, to finish his recon with a nudge in the shoulder.

Enchant glared at the timid and useless pair. "UT, display this problem for me. I want to see what pressing situation delays our glorious return to Traigh."

A hologramatic display of minion#2's view of the UT floated in front of Enchant on the bridge for all to see. It was a view of the stars beyond the gas giant Zoesam. To most, it was nothing more than dots of light from distant star systems, but Enchant saw the threat in an instant; movement, formation – the enemy. "Prepare the fleet, we are under attack."

*

Those aboard the bridge of the Dorsano watched as a cargo port opened up along the side of the Cremmerson.

"Can we get eyes on what's going on there?" said Cooper.

"Already on it," said The Ginge as he lost himself in interface with the UT.

A pulse of fire burst out from the Cremmerson humming close by the Dorsano.

Pete assessed his display. "The Corrax! They've hit the Corrax."

"They OK?" said Cooper.

"There was an explosion... Hard to tell. Stand by."

"What about the shields?"

"Did you see the size of the pulse?"

Cooper took a moment to size up the situation. "Can someone keep me posted on the Corrax?"

"I've got eyes-on," said Rhiannon from her post.

"Thanks Rhi."

Pete turned his attention back to the enemy super ship.

The Ginge narrowed his view in on the side of the Cremmerson. He was greeted with view of dozens and dozens of fighters of various descriptions, pouring out into space and headed their way. "We're under attack!"

"Yeah... thanks," said Cooper.

Hensaro's hologram appeared on the bridge. "Change of plan, attack with everything you've got. Spread the cruisers apart and advance; unleash your attackers to take down theirs!"

"Understood!" said Cooper before turning to Reynolds. "Show time."

Reynolds belted instructions into his communicator while Cooper turned to Knuckles. "Hit them hard."

"Yes, sir!" said Knuckles before he too belted out commands to his team and pulses of fire beamed out form the Dorsano toward the Cremmerson.

"It's turning our way," said Pete.

"I see it. Spread left and try to put the moon between us and them."

"Roger."

Another blast powered past the Dorsano, again striking the Corrax. Whatever resistance the ship gave to the first blast it couldn't match with the second, the front of the craft exploded, opening up like a venus fly trap. Debris spewed into the black.

"Rhi, update?"

"Not good, really, not good."

"Dammit," shouted Cooper as his mind flashed with images of Detarion Anglarae and the faces of the countless people he knew on that craft. He breathed heavily as he panned his next move.

"Hang on, Coops, ships are still launching from the cargo area," said Rhiannon.

"Brilliant! Reynolds, can you make contact? Get them to team up with our fleet."

"Affirmative."

"We're gonna need it," said The Ginge. "There's another wave of fighters coming."

"We'll deal with that when they get here."

"I think we're going to need more than that, there are, like, hundreds of them."

"He's right," said Pete. "This is seriously not good."

"Give me numbers."

Pete consulted his display. "According to the UT, we've got 526 in the air, including those that made it out of the Corrax, they've got... 1224."

"1223!" said Knuckles from behind his turret.

"Keep your fire on the Cremmerson!"

"Just trying to help, jeez."

Cooper poured over the information flooding his way. "Pete, how long until we're out of sight of The Cremmerson?"

"Less than a minute."

"Good. Give us a shout when that happens, Knuckles, when you get the word, have your team switch focus to the attackers."

"Roger."

"Hopefully we can even the odds a little."

Reynolds looked up from his interface displays. "There may another option."

"What have you got?"

"I'm getting word from the Corrax survivors there are still nearly 150 ships sitting in the cargo bay. If we run light here and on the other cruisers we may be able to man those craft."

"Thanks Reynolds. Gandalf can you get comms with Hensaro?"

A hologram of the Fredahn leader reappeared on the bridge. "Cooper. I'm including in Kace Astoy in this communication."

Cooper nodded as the warty Jarden appeared next to the image of Hensaro.

"We are outnumbered and outgunned – suggestions?" said Hensaro.

"We'll soon be out of range from The Cremmerson and we'll concentrate all our fire on the attackers," said Cooper.

"Excellent, we shall head for sanctuary and do the same. Astoy?"

"Our shields are our strength, but we shall follow your lead to cover and concentrate fire on the attackers."

"We believe there may be a large number of spaceworthy craft still docked on the Corrax," said Cooper.

"Acknowledged. For now, let us concentrate our focus on the immediate threat. Get to cover, maintain fire on the Cremmerson, when you are safe switch focus to the attackers."

Kace Astoy and Cooper agreed before the communication was terminated.

"Pete, update?"

"Nearly there... and three, two, one, hang on, a half... erm… a quarter..."

Cooper turned to Knuckles. "Get ready to switch attack."

"Ready."

"...an eighth... a sixteenth..."

"Really Pete?"

"Sorry Coops, I think they're moving."

"Moving?"

"They're banking around, trying to keep us in view, let me ramp things up a bit."

"Make it quick."

*

"Focus your fire on the larger ship," said Isla Enchant. "And keep moving forward."

"High Commander, the other two cruisers will soon be out of range," said Elrus Mersion.

"They will soon get their chance to die. I want them hunted down and destroyed one by one."

"The large ship has strong shields High Commander."

"And we have stronger fire-power. Attack and prevail!" said Enchant as she glared at the battlefield through the viewshield. "What word from Valtress?"

"The outpost is sending all available attackers this way – not long and our force will be doubled."

Enchant eased back into her chair and smiled. "Excellent."

*

"One 128th and... we're clear," said Pete as he made slight adjustments on the controller.

"Good stuff. Knuckles get you team to turn their fire to the attackers," said Cooper.

"Yes sir!"

"First attackers are about to engage the enemy sir," said Reynolds.

Cooper looked at the battlefield display. "Hopefully we can even up the odds a little."

"The Corrax is copping a hammering," said Rhiannon.

Cooper looked at the Corrax and the barrage of fire it was taking. "Gandalf, open comms with Hensaro and Astoy."

The UT obliged and the two holograms were displayed on the bridge in front of Cooper.

"Kace Astoy, can we help?"

"No, we are moving to cover as best we can. You should continue to concentrate your fire on the attackers."

"Agreed – that is the best strategic option right now," said Hensaro.

"It appears my ship has become a valuable decoy – use this as an advantage."

"Once again, Kace Astoy is correct. We continue with our plans. Suggest comms stay open from here on, I think we'll need it."

"Yes," said Cooper, while Astoy nodded agreement.

Hensaro returned the gesture. "Well, what are you waiting for? We have a war to win."

With that, the three captains returned their focus to battle.

"Knuckles – update," yelled Cooper across the bridge.

"It's a shooting gallery out there, Cap. They're quick, but we're still fragging a shit load of them."

Pete rolled his eyes. "Language!"

Cooper looked at the battlespace, now a dance of lights and explosions as the attackers engaged in the space between the advancing cruisers. So much destruction, yet too distant to feel real. He turned to his military commander. "What's going on out there?"

Reynolds didn't lift his eyes from the displays lit up around him. "It's a lot to take in, but at the very least we are holding our own out there."

"We're doing better than that," said Pete. "Loses have been four-to-one their way since we engaged."

"Captain Simpson, we now stand next to you out of sight of the Cremmerson. Diverting our firepower to the enemy attackers," said Hensaro through his hologram likeness.

Cooper smiled and turned his view back to the battle, dozens of new pulses now fired at the enemy attackers advancing on them. "C'mon," he whispered to himself.

*

"High Commander, our attacker craft are taking heavy losses," said Mersion.

"That is not of concern," replied Isla Enchant, eyes not breaking from her view of the battlespace.

"But—"

"You fool; they have sent their fleet of fighters to engage ours. They will continue to engage while their think they can win. Meanwhile, our reinforcements from Tythera are about to show them it is unwise to invest all your resources in one place. They will soon be down to two cruisers, without the protection of their attackers. Victory will be swift."

*

"Erm, Coops," said The Ginge as something new entered his battle view. "Coops!"

"What?"

"We've got a problem."

Cooper looked up from his displays to where The Ginge was pointing out the viewshield. From behind the other side of the moon, hundreds and hundreds of small dots of light moved in formation towards them.

"They're attackers, Coops."

"Oh shit."

Pete shook his head in disbelief.

"Hensaro," said Cooper. "Hensaro, are you seeing that?"

"Yes, Cooper. It is not ideal." Hensaro thought more on the consequences of the new front of fighters. "We are trapped."

***

CHAPTER 30

Cooper looked at Hensaro's likeness. "What do we do?"

The Fredahn paused momentarily in thought. "Kace Astoy, how are you faring?"

"We have underestimated their strength. I fear we may not last long."

"Get to the cover of the moon, my friend," said Hensaro. "It is time to redeploy our numbers. How many capable pilots do you have?

Kace Astoy consulted someone out of view before returning his focus to the conversation at hand. "Not many, possibly 30."

"And how many shuttle craft?"

"We have five craft – capacity to move 245."

"Very well, send a shuttle with your fighters to the Corrax, load up the rest of your crew and ship them to the Themmington or Dorsano."

"Acknowledged. I will remain with those who wish to join me. We will ensure the Narran maintains battle output, while we aim for Hazsis' sanctuary. I fear our goal will remain unachieved."

"Acknowledged, Chief Leader Astoy. You will prevail," said Hensaro. "Cooper Simpson, how many capable pilots aboard the Dorsano?"

Cooper turned to Reynolds.

"It's slim pickings."

"We'll send whatever we've got to the Corrax."

"As will we," said Hensaro. "Let us move light and fast."

Cooper scanned the room, then his mind, searching for the answers. Who should he send? Who could he send? He looked at Reynolds. "Slim pickings – what are we talking about?"

Reynolds kept his smile to himself. The perfect set of conditions had landed in his lap like a pair of Texas Hold'em aces. He acknowledged the moment to himself, breathed deep, then executed. "Most capable pilots are already in the skies. We have a handful of crew from the docks who know the drill, but they're green. The only others worthy are on this very bridge, Sir."

Cooper felt a wave of dread sweep over him, not from what he was about to put himself through, but what he was going to ask of his friends. "Knuckles, Ginge – get your game face on, we're heading to the Corrax. Reynolds—"

"Wait, what?" said Pete.

"You heard Reynolds; they'll be slaughtered if there's no one with experience to help them out. Reynolds, the helm is—"

"With all due respect the best pilot currently on this ship is me," said Pete.

"That's why we need you here."

"But..."

Cooper stood and made his way towards the exit, The Ginge and Knuckles. "No buts, Pete. Reynolds, tell the crew what's going on. Tell them anyone who's up for it to be waiting by the shuttles ASAP."

"Yes, Sir."

"The ship is yours. Look after her for me."

"Yes Sir."

Cooper turned to leave, but he was intercepted by Rhiannon, who tackled him with an embrace. The squeeze pushed air out of his lungs. After taking in her scent and brushing a tear from his eye, he pulled away. They locked eyes. "I've got to go," he said.

"I love you," she said.

"Me too."

With that, Cooper exited the bridge, The Ginge and Knuckles by his side.

*

"They are fleeing the ship, High Commander. It can't have long left now."

"Excellent," said Isla Enchant, glaring at the battlespace, then smiling at the victory she would soon taste.

"Shall we move our attentions to the other cruisers?"

"No! We maintain our current attack until the ship is destroyed. Only then we move on to the next victim. We will crush them all one by one until they are utterly eliminated."

"As you wish."

*

"Easy... easy," said Knuckles.

The Ginge manipulated the controller, steering the shuttle around a tumbling piece of debris as big as the craft he was in. "I got it."

He switched on his communicator and issued instructions to the pilots of the other two shuttles in their fleet. The other vehicles followed his craft around the wreckage.

Cooper entered the cockpit. "How much longer?"

"Well, if we didn't have to dodge around half a star cruiser, we'd be there by now," said The Ginge.

"Well the attackers heading for the Dorsano aren't slowing down."

"Do you think I don't know that?" snapped The Ginge and he and Cooper locked eyes.

"There it is!" said Knuckles.

"I see it!" said The Ginge.

The cargo bay came into view as they neared the rear of the ship. In front of it thousands of shards of metal glowed like fireworks as they spun and reflected sun light or the explosions of battle.

"She looks pretty much intact to me," said Cooper.

"Sweet," said The Ginge. "I'm gonna swing up and over that piece to the left. Looks like we'll have a pretty clean run in from there."

"Good call, I'm going to prep the pilots."

"This probably isn't going to end well for us, is it?" said Knuckles as Cooper was about to head back to the shuttle's cabin.

Cooper paused. "Probably not."

*

Hensaro Althot gazed through the viewshield to the battle that surrounded him and his ship. He searched his mind for the next move, the decision to turn the battle. When nothing presented itself he returned his focus to the other leaders. "Chief Leader Astoy, how are you and the remaining crew holding out?"

Astoy's hologram shuddered as a large blast ripped through the ships defences. "I fear we live in no time now, our fate is sealed."

"Keep fighting, brother, this battle is not over yet."

Hensaro turned to the Dorsano leader, only to see Hank Reynold's standing in his place. "Reynolds, where is Captain Simpson?"

"He has left to lead the attackers in protecting our ships."

Hensaro paused. "Brave, a true leader's decision."

Reynolds nodded.

"I assume you have taken the helm in his absence."

"Correct."

"Tell me; Acting Captain Reynolds, what would you do in a situation like this?"

"Well, sir, we are out-gunned, out-numbered and so far we have been out-manoeuvred. They have rattled us sir, we are in disarray. There would be no shame in leaving now to fight on our terms another day."

Hensaro looked at him and nodded. "It is true. I would consider such a move if we were in a position to do so. Unfortunately, almost our entire fighting fleet are too far away to effect a swift withdrawal."

"Then we call them back?"

"This tactic will only invite the battle to our cruisers as the attackers return. That would not be a good outcome."

"I don't foresee a good outcome if things keep going the way they are."

Hensaro eyed Reynolds. "It appears we are in quite a quandary. I believe fighting is our best hope of survival and our only chance at victory. We will not get this chance again."

"As you say," agreed Reynolds, despite his eyes saying otherwise.

"Maintain fire on the enemy attackers. Let us even up the numbers there. We shall see how confident the Cremmerson is when the odds are more in our favour."

*

The heavy gunship rose above the cargo bay floor of the crippled Corrax, turned towards the exit and fired into space. The Ginge high-fived Knuckles and Cooper before beginning his game of dodge the debris. Out in open space, dozens of attack craft hovered, awaiting orders.

Cooper picked up his communicator. "OK folks, we're going to push out past the junk, bank right and head to the Dorsano. Once we're clear, fan out behind us, get familiar with the controls and ready for battle. Remember, it's just like the sims."

He discarded comms equipment and looked at The Ginge. "How does she feel?"

"Nothing I can't handle." The Ginge navigated around a large piece of fuselage. "Hey, Knuckles, mind shooting me a path forward?"

Knuckles brought up his turret's interface displays. "Thought you'd never ask."

He fired off pulses at the largest offending wreckage, pulverising it in a violent explosion. "Whoa, this is gonna be fun."

The Ginge saw something through the viewshield, leaned in and squinted his eyes. "Can anyone see what that is?"

Knuckles, too, leaned forward. "Not sure, but they're getting bigger.... quickly."

"Captain Simpson, are you there?"

Cooper recognised Dilania's voice over comms. "Receiving. We're heading your way now, won't be too far away."

"Look out!" said The Ginge as the approaching objects raced by them. It turned out to be a trio of shuttle craft. "I hope you didn't take all of the good ships," said Dilania. "If we're going to be good support, we'd better have the right equipment."

Cooper smiled. "That's an affirmative Dilania; I think you'll find plenty to play with."

"I like the sounds of that," said she in a tone way too sultry tone for the occasion, at least for human standards.

*

Isla Enchant watched the continued pummelling of the Narran, while the obscured enemy cruisers fired at her smaller craft. "Mersion! Status of the attackers?"

"The enemy have successfully reduced our advantage, but our flank advances unhindered."

"Excellent, and the cruiser?"

"It surely only has moments left."

"Very pleasing. Enable comms with the raiders."

"As you command."

*

"Captain Althot, The Cremmerson is requesting a communications link, with you and the other captains."

Hensaro looked at the holograms of the other captains, who nodded their approval. "Allow it."

The grey, known as High Commander Isla Enchant appeared in front of them. For a handful of seconds the New Council captains sized up the body language of the grey, while she returned the favour.

"It appears you are as ugly as you are stupid," said Enchant. "At what point in planning this... raid, did you think there would be a favourable outcome?"

"We are here to stop the continued destruction of galactic peace," said Hensaro.

"If your aim is to achieve that through your own destruction, you are doing a fine job." She looked at the three captains once more, taking her time, savouring her superior position. She ended on Kace Astoy. "Why are you still aboard that ship? Surely you are clever enough to understand you only have moments to live?"

"I would rather die here than allow the galaxy to be ruled through force."

"Ahh, yes, I understand the drive of a captain. Do not fear, you'll get your wish very soon."

Even as the last of the words escaped her mouth, the crippled Narran, could take no more. Within seconds of the ship's defences being destroyed, a beam of fire buried deep into the body of the craft, triggering a chain reaction of violent explosions that tore it apart.

As the eruptions of defeat fired around him, Kace Astoy stood in peace, knowing his fate was inevitable. He lowered his head in thought or prayer, his hologram lagging a moment before vanishing from the conversation in a flash.

Isla Enchant laughed. "Don't you find pleasure in the little moments of perfection?"

She glowed at the expression of fear and loss on the other captains, switching her gaze between the two. "The Fredahn or the humanoids from Earth; the ancients smile on me today. Whose pointless death will be next?"

Enchant turned to her crew. "Advance to engage the remaining cruisers, no raiders will leave this system alive."

"It is you who will perish today," said Hensaro.

*

The Ginge steered the gunship low over the surface of the Dorsano. As it breached the horizon, he, Cooper and Knuckles were greeted with the view of a wall of attackers headed their way. "Oh my god!"

Some of the heavy weapons fire from both the Dorsano and the Themmington was now concentrated on the new flanking threat, while the rest served to destroy the Cremmerson's attack craft.

Cooper picked up his commmunicator. "Reynolds, are you there?"

"Captain Simpson. What's the plan?"

"Erm... I'm going for not die right now. I guess we'll punch a hole through their line then pick them off from behind as they head your way."

"Affirmative. Make sure you keep your fighters tight, we've got some heavy fire headed that way at the moment," said Reynolds

"Roger that. Good luck."

"You too."

"OK, you heard the man," said Cooper to his fleet. "Stay tight. We're going to hammer the enemy line, follow us through and then we'll get our own back on the other side."

He looked at the advancing enemy, then at Knuckles. "Give them all you've got."

Knuckles pressed down on the turret and unleashed a heavy attack on the lead raiding ships. Silent explosions littered the black skies as several craft were destroyed, sending shrapnel into other fighters. "Whoo-hooo," he screamed as adrenaline stole his fear.

The Ginge felt it as well, pushing the controller forward, sending the gunship toward the countless enemies. "Ladies and gentlemen, the captain has activated the seatbelt sign – strap in, we are expecting some turbulence."

*

Pete watched the gunship barrel through the enemy line; it was soon lost in the chaos of collisions as the fleet of friendlies followed. He exchanged a fearful glance with Rhiannon, breathed deep then returned his focus to the task at hand. "Captain, we'll have to move soon, the Cremmerson's advancing."

"Roger," said Reynolds as he looked at the hologram of Hensaro. "They're rounding the moon; suggestions?"

"There is only one solution I can think of," said Hensaro.

"Retreat?"

"In part."

"I'm not sure I follow, sir."

"I am going to intercept the Cremmerson."

"Sir? Your defences won't last ten minutes against that thing."

"They won't need to; we'll reach it before then."

"And?"

"And we will not stop when we do."

Reynolds and Hensaro looked at each other in silence. Reynolds thought through the consequences and tried to protest but no words came to him.

"I have asked for volunteers to stay with me; as many as I need to operate and protect the ship. I have ordered everyone else to the Dorsano; will you make preparations to receive them?"

"Of course, I—"

"Once we have passed the attackers I will order all other craft back to your dock. Take them, then go. Understood?"

"Yes Sir," said Reynolds before saluting. "It has been an honour serving with you, Sir."

With that Hensaro turned his attentions to his ship and the tasks at hand.

"You heard him. I need someone to head to the docking bay and prepare – it's going to be mayhem down there."

"I'll go!" said Rhiannon.

"Thanks. Everyone else, I need you thinking clear and acting fast. Stay in communication and let's get as many back alive as we can."

"Yes, sir!"

***

CHAPTER 31

As much as The Ginge squinted his eyes, he couldn't make sense of the view that confronted him. There was no up or down, left or right, just a tunnel of pulse fire, explosions and debris. He jockeyed the gunship as best he could through the mess but the shudder from impact was almost frequent enough to become a constant tremor.

He screamed with everything he had.

Fortunately the ship was strong enough to handle the impact... just. He feared the friendly attackers in his wake might not fare so well.

After what seemed like forever, they breached the far end of the enemy line and pushed out into open space, with a cloud of debris in tow.

"Woo! Nice job you two," said Cooper as he slapped The Ginge on the back. "You can stop screaming now."

"Oh, sorry," said The Ginge as he turned the ship around.

They watched in silence as the wounded fleet of friendly attackers joined them on the other side.

"I can't believe that many made it through," said The Ginge.

"Have you seen the size of the hole we put in the enemy?" said Knuckles.

The Ginge admired the visible circle of destruction, then noticed something else further afield. "Hey, where's the Themmington going?"

The three watching the cruiser head toward the Cremmerson while a flotilla of shuttles and other craft headed from it to the Dorsano.

"That can't be good," said The Ginge.

"What?" said Knuckles.

Cooper picked up his communicator. "Reynolds, you there?"

"Cooper, yes, under the pump, but here."

"What's going on?"

"There's been a change in plan. I'll put you on to Alan for the details. Stay close – updated objectives to follow."

"Roger."

Suddenly Cooper was looking at Alan, sitting in The Ginge's tactician's spot. "What the hell's going on?"

"Hensaro's gone all Jack from Titanic; he's making a suicide run at the Cremmerson."

"No shit!"

"Yep, and get ready, as soon as he passes the other dogfight, Reynolds is calling everyone else back and we're bailing."

"Really?"

"Uh huh, besides which... oh... hang on, looks like you've got company."

Cooper looked up to see two large squadrons peel up from the main enemy fleet and head back toward them.

"Shit, gotta go."

"I'll get turret support on you as—"

Cooper closed the communication and opened up a channel to his fleet. "Stay in formation and on my wings."

*

The Themmington stayed in close line to the moon as it accelerated toward its showdown with the Cremmerson. As it passed the original attacker battle, Hensaro gave the order to retreat and the remaining New Council craft headed towards the Dorsano. The skeleton crew on the Themmington fired on the chasing enemy attackers while Hensaro and his pilot began their preparations for their final assault on the Cremmerson.

*

"All shuttles aboard," said Rhiannon. "We've bunched them up at the back of the loading dock and evacuating everyone now in preparation for the fighters."

"Nice work," said Alan. "Do you think there's enough room?"

"Not sure, how many we talking about?"

"Plenty – a few hundred."

"I guess we're about to find out," said Rhiannon as she closed her communication and ordered the ground crew in. "Listen up, the captain's called the retreat, we're going to need to stack them head to toe in no time flat if we want everyone to get out of here alive.

"There's enough of us to make ten groups of three, let's mark-up territory and start racking and stacking."

As the crew organised itself in to teams, Rhiannon noticed someone else in the cargo area – Jason, Sam and his documentary crew. They'd been filming her. "Not now!"

"We'll stay out of your way."

Rhiannon shook her head in disbelief then returned her focus to the ground crew.

*

The wall of flanking attackers neared the Dorsano. Pete, Alan and most of the crew on the bridge stared, mouths agape, at the sheer volume of enemies. Interface displays and warnings flashed and beeped their disapproval.

"I want every turret firing on them now!" screamed Reynolds, snapping every back to the task at hand.

A wall of pulse fire swept out from the Dorsano, passing the first wave of enemy fire as it went. Dozens of ships were destroyed as the bridge shuddered for the first time.

*

Knuckles sent a plume of pulse fire into the approaching enemy. It took out several craft and scattered the rest. Behind the gunship, several attackers unloaded fire while the enemy restructured and engaged again.

The Ginge swore as the gunship took two direct hits in short succession.

"How we holding up," said Cooper.

"OK... I guess."

"Are we going to be able to get back to the Dorsano?"

"I have no idea."

"Great."

Knuckles neutralised another enemy. "Would you two quit talking about it and start doing it?"

"Have you seen how many bogies are waiting for us?" said The Ginge.

"I'm more concerned about the Dorsano not waiting for us."

The Ginge sighed and accelerated the gunship towards the Dorsano. "Fine!"

Cooper picked up his communicator. "This is it. We're making our run for home. Fire everything you've got... and see you on the Dorsano. Over and out."

As he lowered the device to the console it chirped into life again – it was Hensaro.

"Greetings Cooper."

"Hensaro, I... I don't know what to say."

"There is nothing to say, only decisions made. I simply wanted to thank you."

"Me?"

"Yes, you and your friends make me content to have lived. You've lived up to the reputation that preceded you. You've confirmed to me that life has a value worth fighting for."

"Thank you Hensaro, for everything."

"Goodbye Cooper."

"Goodbye."

*

The first of the original battle attackers reached the docking bay just as the flanking enemy attackers passed over the Dorsano. Several ships reached safe haven before the enemy focused in on the new target. Dual turrets protecting the docking bay flew into life, barraging the enemy attackers, destroying several and steering the rest away from the critical position.

"Alan! Alan, you there?" screamed Rhiannon into her communicator.

"Go ahead Rhi."

"We're not getting anyone home safely if we don't get more turret support to the docking bay. It's mayhem out there."

"Roger that, leave it with me... and stay safe."

"Thanks Alan, you too."

*

"Prepare to fire," said Isla Enchant as the Cremmerson moved into position clear of the moon, opening up line of sight to the Dorsano.

"On my mark... and—"

Between the two craft another appeared – the Themmington. It fired heavy pulse attacks on the larger craft.

"New target – acknowledge."

"Acknowledged," replied the crew before opening fire.

"Give them everything," said Enchant.

"High commander, the enemy wishes to open dialogue."

Enchant scoffed, then nodded and the hologram of Hensaro Althot appeared in front of her. She smiled. "I thought you were foolish before, but this, this is a new level of stupidity. Do you not realise your ship has little chance of prevailing?"

"Life would be boring if we played by the odds."

"Life will be short now you've played against them."

Hensaro laughed. "I already know that."

"Then why are you..." said Isla Enchant as she processed his move and his words. It didn't take her long to draw the only conclusion that made sense. "Fire everything you have! Prepare for evasive action."

*

Rhiannon watched as several of the Dorsano's other turrets concentrated their fire to the skies around the landing dock. They hit the enemy attackers hard, ensuring, as best they could, safe passage for the friendly attackers to reach safety.

"Yes! Go Alan! The geek does good," she said as she pumped her first.

She turned to her crew and barked out instructions as they prepared for the next wave of friendlies.

*

The gunship took another hit and wobbled through the void before The Ginge wrestled it under control. The hull groaned its displeasure at the constant barrage on its structural integrity.

"We good?" said Cooper.

"If you define good as about two hits from exploding, then yes."

"OK, go easy then."

The Ginge looked at the sky teaming with enemy attackers and pulse fire. "That was a joke right?"

"Maybe we should have our exoskins on, just in case."

"Good call."

In front of the gunship, several enemy attackers exploded into quickly expanding debris fields shortly before several friendlies flew through the remains.

"What the?" said The Ginge.

Cooper's communicator bleeped; on the other end, Dilania. "I sense my captain needs me."

"Dilania! Brilliant! Yes, affirmative, help greatfully received."

"Some would expect special treatment for this kind of help," she said as she mowed down another bogey.

"I'll give you a high five when we get back to the Dorsano."

"High five?"

"It's... nothing... just a hand slap gesture."

"Not the two pieces of flesh I would've chosen to connect Captain, but I look forward to it all the same."

The Ginge and Knuckles looked at Cooper, mouths agape, the former mouthing the words, 'Oh my god'. Cooper gave them the she's crazy expression

She shot down another bogey then whooshed past the gunship with her squadron in tow. They arced around then pulled ahead of Cooper's ship. "Follow me."

The Ginge eased the ship into a higher speed, praying the noises wouldn't lead to something worse. "You know Coops; since you're already accounted for I was thinking you could put a good word in with Dilania for me."

"Wow," said Knuckles. "How the mighty have fallen."

The hull groaned again. "Let's just get home first," said Cooper.

*

Reynolds looked at the battlespace holo playing out before him, his focus on the Cremmerson and Themmington. The two craft exchanged heavy fire while they rapidly approached each other. The Cremmerson banked hard while the Themmington zeroed in.

"How long until impact?" said Reynolds.

"Approximately 45 seconds," said Alan.

"And our returning forces?"

"Most of the first group of attackers are back, the second group heading our way now."

"Very well, tell them they have two minutes, we must jump out of this system before further forces arrive."

The bridge went silent, everyone looked at Reynolds.

Pete and Alan exchanged glances, Pete immediately getting on the communicator to Cooper while Alan protested to the acting captain. "Two minutes! There's no way they'll all get back in two minutes!"

"This place will be teeming with enemy cruisers before we know it. We must think of the greater good here."

"But we can't leave Coop—"

Reynolds interrupted Alan by holding up his hand dismissively, he was soon lost in a personal communication. The former sci-fi convention MC knew something wasn't right. He watched as Reynolds listened to the voice on the other end. The only word to come from his mouth were, 'Yes, Sir."

Reynolds closed the communication and glared at Alan. "Well?"

Alan turned defeated from the staring contest, but he knew Reynolds was on his communicator again. He couldn't hear what was said and to who, but it was short and utterly suspicious.

*

Cooper accepted a new communication. "Wassup Pete?"

"You need to get your arses back here now."

"Well, that's the general plan—"

"Reynolds has made the call; we're leaving in two minutes."

"What? There's no way we can all dock in two minutes!"

"I know, something's not righ—"

The gunship shuddered as it took another blast, the strike that would prove its undoing.

*

"Hard left, full power," screamed Isla Enchant as the Themmington grew large in the viewshield.

Around her, the crew and alarms went in overdrive. Collision was imminent. She screamed. The front of the Cremmerson moved up and out of the impact zone, seconds later the Themmington ploughed face first into its belly.

Hensaro's craft tunnelled into the enemy starship until the colliding forces exploded on an epic scale, shooting out a wave of energy.

***

CHAPTER 32

In the docking bay the first attackers from the second squadron started to land. Rhiannon busied herself with her team, frantically directing the increasing flow of fighters. Alan had told her she had less than two minutes.

Attacker after attacker approached. Every time she turned to the exit she prayed the Corrax gunship carrying Cooper would appear. Each time she remained disappointed.

*

Proximity alarms went crazy aboard the bridge of the Dorsano as the crew watched the monstrous collision of the two cruisers. In contrast, the crew stood in silence as the gravity of what they'd witnessed sunk in.

The first to break it was Pete. "Cooper, come in Cooper."

He toggled his communicator. "Knuckles, you there buddy."

And again. "Ginge, do you hear me, over?"

He looked over at Reynolds. "Nothing."

Reynolds locked his vision on the show outside the viewshield. "Woodcock, prepare the crew for immediate departure."

Alan froze, lost in Reynolds' gaze, like a rabbit hypnotised by car headlights, with his mouth agape.

"What?" said Pete. "Didn't you just hear me? Cooper – your captain – is still out there."

Reynolds maintained his gaze on the vast chaotic carcass of battle in the blackness. "Woodcock, prepare the Dorsano for departure!"

Alan gulped as a sickly feeling overwhelmed him; forces were at work far beyond his ability to control. He wasn't sure what was happening; only that, somehow, Reynolds was at the centre of it. Reynolds stared at him, challenging him in a battle of authority, will and inner strength – none of which Alan felt confident winning.

He cleared his throat. "Sir, if we could have another minute or two, there are dozens of additional attackers we could save, we could also find out what happened to Cooper."

"You have your orders, Woodcock."

Again Alan froze; everything was too much to process. "I... I...—"

"You can't just leave them out there!" screamed Pete.

"It is too dangerous, we'll soon be hit with the blast wave from the cruiser collision, it could destroy us. Unless the rest of the Chardrekk fleet get here and do it first."

"Cooper would've stayed for you."

"I'm the Captain—"

"Acting Captain."

For the first time Reynolds turned to face Pete. His eyes were red with rage as he stared down the teen. He maintained his gaze while he barked to Alan through gritted teeth, the tone of his words this time sounding like a threat. "Woodcock, prepare the Dorsano for departure."

Pete and Reynolds remained in a battle of locked eyes when the bridge door opened. A man stepped through, still exhausted from his time aboard the ill-fated GX-124 Narran.

"Captain, I'm here," said Luca Hoffman.

Reynolds smiled. "Excellent. Relieve pilot Bates of his duties."

"Sir?"

"The pilot's chair is yours."

Hoffman looked at Reynolds, then Pete, both still in a staring showdown. He stood halfway between them, unsure what to do, who to disappoint.

Soon, the bridge door opened again and Daniela Rodriquez stepped through, her expression hardened from her time aboard the Themmington. "You requested me?"

"Yes, relieve Woodcock of his duties and prepare the Dorsano for immediate departure."

"Yes, sir."

Rodriquez did not share Hoffman's inability to act. She marched towards Alan, who had inadvertently backed away from his station before she'd arrived. This, in turn, had him totally confused when she sat in the very seat he thought he still occupied.

As Rodriquez began her work, Alan looked at his only friend on the bridge. "Pete?"

Pete threw his hands in the air, defeated, as backed away from the pilot's console. He maintained his glare at Reynolds. "Asshole."

He turned to Alan. "Come on, looks like we'll have to do this ourselves."

The pair left the bridge as the countdown to departure commenced.

*

The sound had been all-encompassing, the kinetic violence of the ship breaking up almost overwhelming. Cooper closed his eyes and screamed, expecting the end.

But the end didn't come.

He was alive. He opened his eyes and quickly realised that may be a temporary arrangement. The large cockpit of the gunship was no more, from where he sat, everything to his right was no longer there. He was spinning out of control in the vast openness of space, his exosuit supplying his oxygen.

He looked to his left; The Ginge was slumped in his chair, cocooned by the small fragment of cockpit that had presumably protected them when their ship exploded.

"Ginge! Ginge!"

No response. He released his safety straps and climbed his way towards his friend. "Ginge!"

He slapped his best friend's face once, then twice. "Ginge!"

Again, no response. He looked around for something – anything – in the remains of the cockpit that could help them. Nothing.

"Ginge!"

This time The Ginge groaned. "Ginge, brilliant," said Cooper as released his friend's safety strap and pulled him clear of his chair.

"Stay with me Ginge, stay with me."

*

Knuckles regained consciousness to find he was drifting alone through space, his exosuit providing protection for the blast, and oxygen to breathe.

He searched the space around him, finally settling his eyes on the mangled wreck of the gunship. He picked up his communicator. "Cooper! Cooper!"

There was a long pause. "Knuckles!"

"Coops! Thank God. Where are you, actually, where am I?"

"We're still in the cockpit, or what's left of it," said Cooper, as he found a smooth part of the twisted fuselage to grip onto then pulled himself and The Ginge to the outside of the hull.

He watched as the view slowly rotated from moon to debris, to Dorsano to more debris, then repeated the pattern.

"I see you!" said Knuckles.

"Where are you?"

"Over here," said Knuckles as he waved frantically.

"What? Where?"

"Hang on; you're spinning out of sight. Look to your left when you come back around."

"Alright," said Cooper.

Beside him, The Ginge started to regain awareness. "What happened?"

"Ginge! No time to explain. We need to spot Knuckles. Knuckles, let us know when you can see us again."

"Will do."

The Ginge took in the scenario. "So, how are we getting back to the Dorsano?"

"Erm... don't know, just need to get the three of us together and have a bit of luck."

"This is probably not the best time to mention it," said Knuckles, "but have you seen what's heading our way from the Themmington explosion? It's doesn't look good."

"What is it?"

"Some sort of blast wave, there's, like, a wall of debris heading this way."

"Awesome," said The Ginge.

"Oh, you're coming into view again," said Knuckles.

"OK, on it, where are you?"

"Here," said Knuckles. "I'm waving."

"I've got nothing. Ginge?"

"Nothing."

"How far away are you?" said Cooper.

"Hard to tell – 80 metres, 100 maybe. Dammit, I think I'm moving further away."

The Ginge watched a piece of debris as it made a fast and close flyby to the piece of hull they now clung to. "Wait a minute, can't you glide in that exoskin?"

"I've tried – nothing. Mustn't work in zero G, or without an atmosphere or something."

Cooper saw the blast wave Knuckles mentioned seconds earlier. "That's really not good."

"Oh shit," said The Ginge.

*

Pete and Alan teleported to the docking bay. Attackers poured in the entry as the ground crew scrambled to load them fast enough so as not to halt the flow. They scanned the space and saw Rhiannon barking out orders and frantically gesturing to the incoming pilots.

They skirted around the edge of the bay to where her team was operating, then sprinted towards her.

"Rhiannon!" cried Pete, Alan in his distant wake. "Rhiannon!"

She turned to face him. She saw the panic and anguish in his eyes, something was wrong. Her thoughts turned to Cooper and her face fell, tears welling at the surface. She breathed out heavily, hoping to expel the negative thoughts, then turned her attentions back to playing Tetris with the incoming attackers. "Where's Cooper?" she yelled to Pete.

"He's… we don't know."

She felt her soul drop through her feet. The thought of losing him too much to contemplate, she turned back to ushering the next attacker to safety. "Ginge and Knuckles?"

"We're not getting comms with any of them."

"We sending out someone to get them?"

Alan huffed in next to Pete then doubled over, a wheezing wreck. "Reynolds… kicked… us… out."

"That's why we're here," continued Pete. "We're gonna go after them. Can you get us a ship?"

"Attention!" came a voice through the speaker system. "Captain has ordered our departure. Clear the bay and prepare for hyperspace. Repeat, clear the bay and prepare for hyperspace."

Warning sirens and flashing lights bleared out before the announcement was made again. At the same time the large to the loading bay lurked into gear, slowly closing.

"What are they doing?" screamed Rhiannon. "We've still got people coming in!"

"It's Reynolds," said Alan, slowing finding his way to upright. "He's gone cray-cray."

Rhiannon gazed on in confusion.

"Bottom line," said Pete. "If anyone's going to get them it's got to be us."

Rhiannon paused for a second or two, while she played out the options in her mind. She looked at the docking gates, narrowing by the second, then at the ships already parked and abandoned, finally at the attackers awaiting her direction. She turned to another member of the ground crew. "Stephens! Take over for me."

"Yes, Sir."

She signalled Pete and Alan to follow as she made her way in and under the parked attack fighters. "There's a rescue ship this way."

Pete could see the target craft sitting head and shoulders above the attackers. "Rhi, they're packed in like sardines. How are we going to get it out?"

"You're the pilot, you tell me."

*

"I see him!" said The Ginge.

"Where," replied Cooper as he scanned the general direction of The Ginge's pointed finger.

"Do you see the three tumbling silvery bits? Just above that and to the right. See? See! Wave again Knuckles. Right there, see?"

"Nuh."

"Guys, you're still getting smaller, and that blast is getting a whole lot bigger," said Knuckles. "Anyone got any ideas?"

"Have you got anything you can use as propulsion?" said The Ginge.

"Pro…what?"

"Propulsion, you know, like a rocket."

"You're joking, right? Because I totally forgot to wear my personal rocket pack today!"

"Doesn't have to be a rocket numbnuts, anything you can fire off behind you will send you forward – a gas or something?"

"There's only one thing I can think of that fits the criteria."

It took a couple of seconds for Cooper and The Ginge to join the dots. "Eww."

"Unless you've got any better ideas, I'm going to give it a crack."

"Crack, really?" said The Ginge.

Through the audio connection, Cooper and The Ginge heard Knuckles hold his breath before letting out a groan. Seconds later he repeated the sequence.

"Umm, dude, can you switch you mic off or something?" said The Ginge.

"I think I got some movement."

"…movement now? Wow!"

Knuckles gritted his teeth and grunted again.

"I don't feel so well," said The Ginge.

Cooper scanned the skies, trying to spot something, anything, that could help them. "Knock it off!"

"I'm serious; I think I can smell it."

"Knuckles, is it working or not?" snapped Cooper.

"I think so… maybe… actually, it's hard to tell."

"If it's not, move on. We need to find something and fast."

"Alright. I think I've got enough for one final assault."

There was silence at the other end as Knuckles put all of his concentration and energy into what he hoped would be a triumphant trumpet for travel. His success was so emphatic it could be heard through his audio connection.

Again his communication was greeted with an extended pause.

"Are you OK?" said The Ginge eventually. "That sounded wet."

"I, erm, may have… overcommitted."

"Oh, wow."

"Are you moving forward?" said Cooper.

"I'm trying to move as little as possible right now to be honest."

"I mean, did it work? Are you headed our way?"

"Umm… ermm… no."

"Well there's a minute of life we'll never get back," said The Ginge.

"And we won't have many more if we don't find another way out of this."

*

Pete jumped behind the controls of the rescue craft, while Rhiannon and Alan found seats to either side. He looked out over the docking bay – a scene of chaos as the tail end of the attacker's fleet took their chances with the narrowing entryway. "Strap yourselves in, this could get ugly," he said.

"We going to make it?" said Alan.

"Maybe."

Pete tangled with the controller and sent the ship skyward until one of their landing stands became entangled in that of an attack ship's. The rescue craft jolted forward, pushing the cabin crew with it.

"Everything alright?" said Rhiannon.

Pete jostled with the controller, nudging the craft in little increments in an attempt to release it from its prison. "Stupid thing… stuck."

Rhiannon's eyes remain fixed on the doors. "Hurry, Pete!"

He dropped the craft back down and rotated it around to the left before slowly raising it again. Just as he thought he was about to move clear the leg stand hit the snag once more. "Oh, come on!"

"Pete!"

"I know, I know."

Pete repeated his move, this time rotating the craft right, but his efforts only yielded a similar result.

"Pete, please!"

"Screw this," he said as he thrust the ship upwards, adding more and more power. The craft groaned and complained as the forces pulled at the structure of the leg stand. The engine roared. Pete screamed. "C'mon!"

Then something in the entanglement snapped, shooting the ship up at speed. Pete continued his scream as he reversed the thrust and slowed the craft before it hit the docking bay ceiling like a pickle hurled from a teenager's burger at a fast food restaurant.

"Go, go, go," shouted Rhiannon.

Just as Pete punched the craft forward, two incoming friendlies raced towards the ever-disappearing gap between the exterior doors. They clipped wings, sending the first craft ploughing into the parked vehicles below, while the second met with one of the doors. The doors seemed to go into a pre-programmed emergency mode, accelerating their efforts to close.

"Oh no," said Rhiannon as another attacker whooshed through the gap, followed by another, the latter needing to rotate sideways to fit.

A third craft was not so lucky, the gap between the doors too small to pass. Only the cockpit reached safety, before the main hull struck the sides of both doors, stopping in its tracks. The cockpit section ripped from the body, and ploughed into the parked attackers already ablaze from the previous crash.

"Pete!" screamed Rhiannon and Alan as they neared the impenetrable doors.

The pilot reversed thrust and the engines screamed again. He braced for a collision that never came as the craft pulled to a stop just in time.

The rescue craft hovered as Pete caught his breath. To his side, Rhiannon burst into tears.

The gravity of what just happened hit them all. They were stuck, their friends were lost.

Pete turned to Rhiannon. "I'm… I'm sorry."

Rhiannon released her harness and moved to where Pete sat, then buried herself in his arms and wailed.

Alan came in to join them. "Shh. It'll be alright," he said in soothing tones.

"Alright?" said Rhiannon. "They're dead."

"Hey, we don't know that. This isn't over, not by a long way.

***

CHAPTER 33

A wall of ship carcasses and the debris of war became caught up in the blast wave that pulsed out from the starship collision. The junk pushed out in every direction. Some showered the moon of Hazsis in a blazing show of meteorites, while the rest pushed into the infinite black in every other direction.

On a small piece of wreckage, Cooper and The Ginge watched the hellish torrent approach. To their left the Dorsano began manoeuvring up and out, avoiding the denser areas of destruction and preparing to depart.

"I'm starting to think this isn't going to end well," said The Ginge, watching the last ride out of enemy territory leave without them.

"I think you might be right," said Cooper as he engaged his communicator. "Knuckles, you still with us?"

"Still here," came the response, broken and digitised.

"You're breaking up a bit there, man. Hang in there, alright?"

Cooper's communicator barked out once more, this time nothing was audible. The interference from the masses of wreckage between them and Knuckles were now too big an obstacle to overcome.

"I can't even see him anymore," said The Ginge.

"Short of trying to push off from here, leaping a few kilometres through space, avoiding the wreckage and landing on the side of the Dorsano, is there anything else you can think of?"

The Ginge laughed; full of defeat. "I think we'd have a better chance jumping towards to the moon."

'True." Cooper turned to face satellite. "Actually, that could work!"

"What?"

"Think about it, the exoskins would kick in when we reach the atmosphere and we could glide to surface. We'd be shielded from the blast wave too."

"Coops that's totally insane. We get one move; once we push off from here we have no control at all. Even if we did nail our aim, like absolutely perfectly, there's no way we'd dodge all the junk on the way."

"Have you got a better plan?"

The Ginge thought on the problem eventually conceding defeat with a shake of his head.

"Exactly. I say we go for it."

"It's suicide, for sure."

"It's giving us a chance. Staying here is suicide."

Cooper negotiated the wreckage they were perched on and got himself into a position where he could propel himself forward. "Ready?"

"You're insane," said The Ginge before joining him.

*

"Cooper? Coops!" said Knuckles as he hit his communicator with his other hand. "Ginge? Anybody?"

He drifted further through the void. A piece of metal zoomed past him at incredible speed, missing him by millimetres. In the distance he saw the Dorsano preparing to retreat. Further out, the blast wave that would likely finish him continued its unrelenting approach.

He looked around for something to help; a way out. There was nothing. "Cooper!" he said into the communicator once again.

Again, no response.

He breathed heavily as his fate became crystal clear. While nothing could save his body, he knew there was still hope for his soul. He sniffed back the emotion and engaged his communicator.

"To my sons. If you find this message, know that your dad… loved you. I may not have been there for you all the time but you never left me. My time has come and all I want is to be with you. Take care of yourselves, each other and your family."

He sniffed again.

"You are different than anyone who has come before you. Don't shy away from that difference, don't be ashamed, it is the thing that makes you, you.

"Balart, Fjard, Michael; I love you.

"Wendy Wendy, look after them, huh?"

*

Pete negotiated the comet to some of the limited remaining space in the loading dock and touched down once more. He planted the ship on the surface and he, Rhiannon and Alan rushed to the exit.

The hatch opened and the three jumped out into the chaos of alerts and alarms as fire raged and the ground crew rushed to douse the flames and save the injured. The panic, movement and noise conspired to hide the message that scored the drama, repeating through the sound system, telling all of the impending hyperspace jump.

Rhiannon hit the deck first, Pete soon by her side. She scanned the scene and tried to focus her mind. Before she could grasp answers, Jason, Sam and the documentary crew appeared, training two cameras on them. She groaned her frustration.

"Seriously Jason?" said Pete. "I'd expect it from Sam but not you."

Alan caught up to the pair and the paparazzi. "What's… the… plan?"

Rhiannon picked up her communicator and pinged Cooper – nothing. She repeated the call, crying her partner's name into the device.

As the camera lenses pressed in on them, Pete put his arm around her. He wrapped her in a consoling embrace, obscuring her face from the camera's view. He pushed at the camera lens then glared at Jason and Sam.

"He's alive, Pete, I know it," said Rhiannon.

"I know, I know," replied Pete, unsure whether he believed the words.

"Attention, ground crew, departure imminent. Clear the bay and prepare for hyperjump," came the voice over the sound system.

Alan jumped between Rhiannon and Pete. "Come on, we have to go!"

He headed towards the main entrance, pulling the others along. After a step or two they followed of their own accord. As Pete passed Sam he physically threatened to punch him. Sam braced for an impact that never came. Pete repeated the threat with Jason and when Jason flinched Pete kicked him in the shin.

Soon, Rhiannon and Pete hit top speed, passing Alan as they raced for the exit. Rhiannon engaged her communicator once more, this time pinging the bridge.

*

"Captain Reynolds, it's Rhiannon," said Rodriquez.

"Display," he said coldly.

Rhiannon's hologram entered the bridge and Reynolds looked her up and down. "What is it?"

"Captain, you need to delay departure and reopen the docking bay doors. There are a number of attack—"

"I'm afraid that is not possible."

"Of course it's possible, just—"

"You're best served preparing for departure, we're—"

"Cooper's out there."

Reynolds eyed Rhiannon's hologram and considered his words. He knew nothing he could say would justify his course of action to her, nor did he have the time or inclination to do so. Instead, he tapped at the digital interface in front of him and her communication was terminated.

He turned to his pilot. "Hoffman, update?"

"Sir, I don't think we're going to get the clear air we need for safe departure before the blast wave hits."

"We have one minute twenty seconds before the blast hits. Do your best to get us clear. We depart in one minute. Rodriguez, update the crew."

She exchanged glances with Hoffman. "Sir, the docking bay is ablaze and the crew are fighting it as we speak. Leaving now would be dangerous."

Reynolds stared at Rodriguez, then Hoffman. "We are departing in 50 seconds. Update the crew."

"Sir, yes, sir."

*

"Get in here, I'll teleport us to my quarters for the jump," screamed Pete over the noise.

Rhiannon and Pete huddled in tight, while Pete interfaced with the UT. "Dark A$$a$$in, you heard the plan – execute the teleport."

"Your teleport capabilities have been restricted," said Pete's UT avatar.

"What? Why?"

"Acting Captain, Hank Reynolds has denied you—"

"Douche," screamed Pete as he punched the closest attack vessel out of frustration. "Arrgghh!" he screamed as his hand throbbed.

"We've got to go, now!" yelled Rhiannon as she sprinted to the exit doors, the others in tow.

The man door to the docking bay opened into a large walkway heading to the centre of the craft. As Pete, Rhiannon and Alan passed through, a large shield door from above began to lower.

"Attention crew, hyperspace jump in 50 seconds," came another announcement.

The trio continued at speed down the corridor. On each of its sides, crew members lined up, backs to the walls, strapped into place and prepared for the hyperspace. They barked encouragement to the trio and the other escaping crew, urging them to hurry.

"Attention, crew, hyperspace jump in 40 seconds."

Pete's body burned as he put his head down and ran. He felt the crew's energy spur him to feats of speed he didn't think he was capable of. But, through the soundscape of craziness and a body full of adrenaline one crystal-clear thought emerged – the nearest available harnesses were still some distance down the corridor, he would only just make it. And if he only just made it, Rhiannon was in trouble and there was no hope for Alan.

He stopped and turned. He was surprised to see Rhiannon right on his tail.

"What are you doing?" she said.

"Keep going!" he screamed.

"Attention, crew, hyperspace jump in 30 seconds."

Pete closed his eyes and summoned the UT as other crew passed in their desperate scramble for safety. He requested an exoskin and was duly fitted with the invisible upgrade. He adjusted to the feeling and willed his mind to elevate. He took to the air and headed towards Alan.

*

Alan wheezed and hobbled forward, sweat dripping from his forehead as the unfamiliar pain of physical activity assaulted every body part – internal and external. He looked up at the long and improbable task ahead, roared then pushed his body through the torture.

Everything beyond the thumping of his heart and endless stretch of crew-lined corridor ahead became background noise. Other crew members passed him left and right. In front of him someone tripped and another fell over the fallen frame of the first. He had no time to dodge, he did the only thing he could think of – he jumped.

At least that was his intent. Betrayed by his weakened body, he barely left the ground and collided solidly with the first fallen man. He tumbled forward then, something unexpected happened. He rose upwards.

"Attention, crew, hyperspace jump in 20 seconds."

Pete gritted his teeth as he wrapped both arms around the portly MC and tore through the air. He could barely reach around Alan's waist and he had to interlock his fingers to ensure they stayed in place.

"Oh my god, dude, you're heavier than I thought."

They flew over the other escaping crew, Pete barely in control of his flight without the ability to use his arms for balance and the burdensome load he gripped.

"Broad shoulders," defended Alan.

"Broad everything! Brace yourself!"

As inelegant as the flight was, the landing was worse. Pete slowed as best he could and tried to hit the ground upright. All was going to plan until Alan's feet hit the ground. He started his feet in a running pattern just before they landed. But, like trying to jump onto a moving treadmill, the task required a level of strength, power and experience he did not have. His feet became tangled and he tumbled forward, sending Pete over the top of him.

Alan face-planted the deck while Pete army-rolled then tumbled to a stop.

"Attention crew, hyperspace jump in 10 seconds."

Rhiannon was just reaching the pod when the crash happened in front of her. She rushed to Alan. Pete, too, recovered from the fall, saw Alan in trouble and rushed to help.

"Alan! Get up!" screamed Rhiannon as she grabbed his arm and pulled.

Alan groaned.

"Alan!" Pete grabbed his other arm and the pair lifted his upper body to a sitting position.

Alan was dazed and unresponsive. Pete and Rhiannon tried to drag him to the side of the corridor, but he wouldn't budge.

Pete looked at Rhiannon. "Get in the harness, I've got this."

Rhiannon glared back at Pete. "No, you need my—"

"Go!" said Pete as he grabbed Alan by the waist and engaged his exoskin again.

Rhiannon watched the two rise above her, and reached up to help keep them steady.

Then Pete charged forward through the air at the wall, and struck it at speed. Alan, back to the wall, took the brunt of the impact. The strapping on the wall recognised the human form and sent shoulder, waist and ankle restraints into action, pinning Alan into place.

Then he felt it, then gut-churning sensation of a hyperjump. The place where senses could no longer be trusted and logic left. He turned to Rhiannon to see her getting strapped into safety. They stared at each other as they realised one was safe and the other not.

"Pete," she screamed. "Pete!"

***

CHAPTER 34

Cooper and The Ginge watched the Dorsano disappear from sight – an unnecessary exclamation mark on the cold reality of their plight.

"Hey Ginge," said Cooper as he prepared himself to jump. "In case this doesn't work—"

"You actually think there's a chance it will?"

"...I just wanted to say, with everything we've been through. I... I... love you, man."
The Ginge scoffed.

"I'm serious."
The Ginge looked at him.

"I'm just saying you're my best friend and I just wanted you to know that."

"Thanks man," said The Ginge before an awkward pause. "I... I... me too."

They stared at the distant moon and the twinkling debris field in front of it. They stared at their mortality.

"Ready?" Cooper said eventually.

The Ginge gulped. "Ready."

"Where do you think you're going, my captain?" came a familiar voice through the communicator.

From under them, a gunship rose. Soon they came face-to-face with Dilania and her alien crew.

Cooper couldn't speak for several seconds, as he processed what was happening. "Wha... Why aren't you on the Dorsano?"

"It seems the acting captain isn't one to wait around."

"He left without you?"

"More accurately, I believe he left without you. My crew and I seem to have been caught in the politics of it all. Since we had nowhere else to go, we thought we'd see if the ship's real captain was around somewhere."

"Thanks Dilania, we owe you... like, everything."

"That is good information to know. For now though, you might be better served on the inside of this ship. We still have a long way to go to be safe. There's an airlock on the underside of the craft. Bringing it around now."

The ship manoeuvred into position with a few subtle jumps. When the airlock opened, Cooper and The Ginge climbed to relative safety.

Once inside, the pair interfaced with the UT and released their exoskins. They followed a big blue frame to the cockpit where Dilania was at the helm.

"Captain Simpson, The Ginge, you remember Gluff Hurn from the Dorsano."

The humans nodded to their silent blue guide then turned their attentions to the three short, skinny aliens manning various drive stations.

"And, of course my little Kolprai helpers Edson Pelk, Gwii Starn and Fillopil Vargreet."

The humans and the three Kolprai exchanged nods.

"Thank you all," said Cooper. "So, what's the plan?"

"Most options are beyond us right now. Our most imminent threat is the blast wave, and we're expecting additional forces from Traigh any minute. The moon has an atmosphere in the tolerance for all species here. That is our target."

"That's where we were going too."

"In what craft?"

"We thought we could jump there."

There was a pause before the three Kolprai started laughing, then Gluff and Dilania joined in.

"See, I told you it was suicide," said The Ginge to Cooper before joining in the laughter.

"With greatest respect to my former captain, it speaks volumes of your courage… not as much for your understanding of physics."

Cooper felt himself going red. "It's not like we had much of a choice at the time."

"Come," said Dilania in a tone forceful enough to stop the mirth. "We can laugh at Earth's leader more once we reach Hazsis."

"Wait," said Cooper. "Knuckles is out there somewhere, we need to get him too."

"Universal, search the debris field for the Earth human known as Bradley Thomas."

Knuckles' likeness was displayed among a number of interfaces floating in front of Dilania and her team. "This being is not in the debris field."

"Yes he is," said Cooper. "We saw him a couple of minutes ago."

"He could've made the Dorsano?" said The Ginge, in a tone that was pure hope and zero belief.

"Captain Duritee, our window to reach the cover of Hazsis before the blast wave hits closes, we must depart!" said Edson.

Dilania looked at Edson. "Do you have any idea what Knuckles has done for this galaxy? We give the humans all the time they need."

Cooper nodded to Dilania. "Gandalf, have you any idea where Knuckles might be?"

*

Knuckles started to panic. He didn't realise what was happening at first as his breaths started to become faster and shallower and his heart thh-thhnnked through his chest. He turned to the Dorsano only to see it spew out a show of light before disappearing into the void.

His heart raced faster. He started to feel light-headed and gasped for air.

"Lilly," he wheezed to his universal translator, "how much oxygen has this thing got?"

"The exoskin's oxygen harvesting and regenerating systems will still give you another 14.85 minutes' worth of breathable air."

He tried to let the words settle him and let his breathing find a rhythm. While the sensation that felt like a hand around his windpipe released its grip a little, it did not let go. He took in some more meaningful oxygen while he could. This was it – his last chance to live… to see his boys again.

"So, what you're telling me is the blast wave is my biggest problem right now?"

"That will reach you well before you are in danger of running out of oxygen."

"How long have I got?"

"Two minutes, twenty three seconds."

"That's just great."

He tried to focus his mind on something – anything he could do to protect himself from blast. He looked at the objects floating around him. "Hey Lilly, you see that piece of hull, with the red alien writing on it?"

"Yes."

"If I could get over to it and use it as a shield, would it protect me from the blast?"

"That is a fragment of hull from a Chardrekk gunship. Using it as a shield would drastically increase you survival chances."

"Sweet, any ideas on how I can get across there?"

"It lies 23m from your current location and is moving further away. You would need to jettison a weight in the opposite direction but you do not have anything ample enough to propel you there. Alternatively, you could also discharge a gas—"

"Forget it! I'm out of ammo."

"I do not understand."

"And I'm OK with that."

Thwack. An object ploughed into his shoulder sending a shot of pain through his unprepared body. It also pushed him into a spin. Suddenly his view changed from the carnage of battle, the blast wave and moon, to the view of the gas giant Zoesam. It was as if the planet's striking, spiralled orangey-red surface hulked over the remains of war, either burning with disapproval or excited by the carnage.

Then he could only see stars – a vista of glowing pinpricks, before his view turned back to the battlespace, moon and blast wave.

"Ouch! Just for future reference Lilly, I like a bit of a heads up when I'm about to get smashed by space junk – it's just a rule I have."

"Apologies, the battle has caused connectivity problems, I am operating on local feedback only."

"What are you telling me?"

"I know nothing more than you."

"What about ideas – can you think of something I haven't? To help me get through this."

"Even if you were to find a way to the gunship hull, your survival chances are minimal."

"I thought you said they were greatly increased."

"And they would be, several fold."

"So, I'm screwed either way?"

"If, by screwed, you mean unlikely to survive, then yes."

"Awesome."

He watched the orangey-red of the planet Zoesam pass by him again. He noticed he was breathing normally. Instead of searching for a way out, he took in the world's harsh beauty. Instead of thinking forward, he started thinking back. The life he'd lived, the things he'd done, the family he'd created – he was at ease.

Moisture formed in his eyes – driven there through pride.

"Lilly… thanks."

"I have failed to find a solution for survival."

"I mean for everything."

He closed his eyes and transported himself to Gartorgia, sitting at the dinner table, his boys throwing food, yelling and generally running amok. He smiled.

Then a beam of light hit him, so powerful it passed right through his eyelids. He open his eyes then instinctively used his hands as a shield from the blue luminosity. Then he felt his body move, dragged toward the light.

"Lilly, what's happening?"

"A ship has arrived; you are being pulled towards it."

"Ship? What ship?"

"For someone who considers himself far more intelligent than his family, why is it that we are constantly smart enough to be there to save you?" came a voice through Knuckles' communicator.

"Vangillor, is that you?"

"The one and only. What other fool do you think would risk his life for yours?"

Knuckles let the tears in his eyes consume him, turning into a torrent.

"Is our rescuing you from certain death displeasing?"

"I'm sorry," said Knuckles. "No, it's… thank you."

"Again?"

"Yes, again."

***

CHAPTER 35

Pete floated through the air as if in a dream – his vision hazy at best, a crazy acid trip at worst and sounds, well, they were just random muffled illusions without any pattern or context to pin them down.

This was hyperspace travel and he was free-forming his way through it unrestrained and unprepared. The take-off had been crazy enough and, although the effects were as much psychological as they were physical, he definitely didn't want to experience the landing in a similar fashion.

Around him he could glean enough from his environment to make out the row of crew in their restraints, lining the corridor wall nearest him. Then Rhiannon said something he didn't quite make out. At least he thought those things happened.

He felt something collide with him. It made him jump. Another aimless, unrestrained crew member, perhaps – he couldn't be sure.

The only thing he could be sure of is he needed to find an unused harness, stat.

He guessed a direction, guessed a way to move and hurled himself into the crazy. He screamed – it sounded like an underwater conversation, played backwards through a voice modulator.

He floated through an ethereal mind-bend that replaced the corridor, without any perspective on how, or indeed if, he was floating through the space. Until he hit something. Hard. Hard and flat, definitely hard and flat.

Was it the wall? He reached out for it but the mystery object was no longer within his reach.

Then his ribs screamed as another collision occurred, this time with an object colliding with him. It seemed to have more give than the previous contact – another crew member maybe.

Pete started spinning and tumbling out of control. Any mild sense of perspective he had was gone. Colour and noise echoed around him. He stuck out his hands both for protection and in hope of gripping something.

Then he felt the wall again. His hand pressed against it hard, then hard as he rode up on it. Soon his arm bent back at an unnatural angle. He felt a crunch in his shoulder, then a screaming pain.

He screamed back; taking on pain's challenge. He kept his fingers sliding along the wall; searching. Then he felt it – a restraint. He snatched his hand closed around the object. The momentum took his body out from the wall and over. All the force challenging the strength of his burning shoulder. He screamed again and prayed he could hold on.

His body rag-dolled around his shoulder pivot point until his head was thrown into the wall. Through the pain he maintained his grip.

He pulsed some heavily deep breaths in preparation for the pain he was about to receive, then emptied his lungs and pulled his body towards the restraint via his injured arm.

It was only when his face was pressed up against the strap; he could actually confirm it was what he thought it was. It was just the perspective he needed, he saw and felt a tangible object – he had orientation. He fished around for the other strap with his good hand, gripped it and used his new connection to spin around.

He slipped his good arm in the strap, pressed his back up to the wall and delicately eased his bad hand in. When the restraint detected him in position, the straps closed in around him, cradling him to the wall.

He was safe.

*

Dilania steered the gunship clear of the last of the heavy debris and soon the moon of Hazsis was within reach. It was a ball of green and yellow vegetation, dotted with lakes and rocky peaks in equal measure. Although it lay directly in front of her, she skirted the craft to the side, rotating around its surface. It was the only way they could land and avoid the heart of the blast wave.

She consulted her readouts. "I'm going to make landfall over the forested area, near the base of the mountain range in sector… 128. That should keep us out of the major fallout. Edson, identify a location for shelter."

"Affirmative captain," said Edson as he and he cohorts called up further terrain details and went to work.

Just as they did, the atmosphere underneath them radiated with blinding light as first one, then several, pieces of battle-carcass, crashed into the atmosphere and towards to surface.

"Holy crap," said The Ginge.

Then the blast wave hit the atmosphere in all its brutal beauty, showering Hazsis with a bombardment of meteorites that dwarfed what they'd just witnessed.

"It's going to be safe down there, right?" said The Ginge.

"Safer than anywhere up here once that blast has passed and the Chardrekk can move their cruisers in," said Dilania.

"Good point."

"Captain, we've located a series of caves that should provide adequate protection. Marking on your displays now," said Edson.

A red sphere pulsed on the display where the mountain range met the forest.

"Excellent," said Dilania as she manipulated the controls.

Proximity warnings began echoing through the cockpit as the outer fringes of the blast wave came dangerously close to the ship.

She looked at Cooper and The Ginge. "You might want to prepare yourself; this will not be a comfortable entry."

Just as the two were ensuring they were appropriately harnessed, the ship darted down and to the right. They lurched around and struggled into their restraints as the ship made another sharp turn.

They hit the atmosphere and Dilania toggled the controls left and right to get a feel for the ship's handling under the local conditions. She eased off her descent to do the same, before being satisfied she was in familiar territory and diving the ship hard towards the surface.

Cooper and The Ginge screamed as they were pressed back into their seats, watching the action through the viewshield as if on a rollercoaster barrelling through fireworks.

The horror plummet lasted nearly a minute – an agonisingly long minute. The Ginge maintained his scream through the entire descent, ping-ponging his focus between the meteor-filled sky, the rapidly approaching surface and Dilania dancing her hands around the ships controls.

Cooper meanwhile found an acceptant calm. He sat and watched the entire event unfold as if watching a television. He had lost his partner, his ship and his captaincy. He had been betrayed... by someone from his own world. He had sent many humans into a battle that would kill them; Hensaro and many others had sacrificed themselves, and for what? With all the loss he'd witnessed and the hollowness that echoed around where his soul once was, well, he was beyond caring whether they made it surfaceside.

The forest and cliff neared. If the passengers of the catapulting ship had cared to look, they would have noticed flora of true magnificence soaring high into the air. Flying creatures soared above the upper canopy of vibrant rainforest and, in a few spots on their right, fire rimmed the edge of the craters torn into the foliage.

Dilania pulled the craft out of its plummet, levelling off just above the tree tops. In front of them the sheer face of a large rocky range called to her. Seconds later another small fragment of battle junk hammered into the forest, tearing century-old trees apart.

"Directly ahead, Captain," said Edson, consulting his displays before looking out the viewshield. "The entrance must be under the tree line."

Dilania pulled the gunship to a hover near the target point. "Gluff, fire on the target marker."

"Yes captain," said Gluff as he sent a volley of pulse fire at the trees that guarded the cliff face.

Within a few seconds, several large trees splintered to the ground, revealing a cave entrance, easily large enough to handle the gunship.

"Good work," said Dilania as she eased the gunship forward.

As they reached the cave entrance, she engaged the UT and the gunship lit up the rocky hollow.

Edson consulted his readouts. "It appears geologically stable, captain. I'm getting some lifeform readings but—"

The rock face in front of them shuddered as a large piece of space junk crunched into the cliff high above. A plume of rocks and dust exploded out and over the gunship.

Dilania jetted the craft forward into the cave – the medium-sized vehicle needed skilled hands to navigate the tight squeeze.

Over the next minute the cave, the ship – everything – vibrated as sections of cliff avalanched to the forest floor. Soon, the light beaming in from the entrance was no more, as the entire space was blocked with rubble. They were left with the gunship's exterior lights, bathing the rocky walls in yellow. A wall of dust engulfed the ship – the yellows became browns, the cave walls barely visible.

Dilania gently lowered the gunship down. "That should make eluding the enemy easier than first thought."

"What, we're staying here?" said The Ginge. "For how long?"

"As long as we can," said Dilania. "Assuming the cave structure still stable – Edson?"

He consulted his displays. "Affirmative. The entrance will maintain its integrity. The rubble can be easily removed when we are ready. Alternatively, this cave system is vast and complex; it appears other exits may be present."

"Thank you Edson. Captain Simpson, once the worst of the conditions pass us, I suspect the Chardrekk will arrive in large numbers to sweep for survivors. We have found ourselves blessed with an ideal location to remain out of the enemy's gaze. I suggest we hide here until it is safer out there. We have enough food supplies to last several days and it will give us time to repair the craft. Additionally, stealing an enemy ship will be far easier when they are no longer hunting for us."

"What do you mean steal a ship?" said The Ginge.

"It is our only way out of this system."

"I thought you said you were repairing this ship."

"This is a gunship; it is not capable of intersystem travel."

The Ginge opened his mouth but realised he'd run out of things to say. Instead he sighed and looked at Cooper for answers.

"Sounds good to me," said Cooper.

The Ginge sighed again.

*

Knuckles limped into the cockpit of the Gartorgian craft and looked at his brothers-in-alien-law. Vangillor sat at the helm, studying the human, a smile glued to his face. He eased back into his captain's chair. "It's hard to know whether to admire your bravery or laugh at your stupidity."

He took in a suspicious breath. "And what is that smell?"

Knuckles gathered all the pride he could muster. Instead of responding he stared at Vangillor with his chin up.

"Every family has a weakest member, but they are still family," said Vangillor.

"You didn't have to come here."

"But I did. Just as we did near your home world seven seasons ago." He gave the human a pitying look up and down. "You are bleeding."

"I'm fine."

"You are in pain, you smell of excrement and until two minutes ago you were floating to a sure death in enemy territory – that is a strange definition of fine."

Knuckles examined the pain in his shoulder with his hand; he flinched through gritted teeth. He knew he must have looked a disaster and smelled worse, but none of that bothered him. No, his pain ran far deeper than his appearance. He was headed back to Gartorgia, his reputation, should he still have one, gone. Cooper and The Ginge were most certainly dead and any chance of being able to atone for their loss was now impossible. How could he ever go back to Earth now? What would even be left for him if he did?

Like the fate he feared for his children, he felt lost – parented and orphaned by two separate worlds.

"On our world it is customary to show some gratitude when someone saves your life."

Knuckles could feel the edges of his soul hardening, so he could feel nothing else. He looked Vangillor hard in the eye. "Thank you."

Vangillor laughed, then looked around the cockpit at his brothers. "There you go, the human thanks us; our mission has value."

He hammered the controls then looked at his brother. "Regilor. Prepare for departure."

*

The haze of confusion lifted over several jarring seconds as the Dorsano, eased into a quiet distant orbit over Baltwae. Wails of pain echoed up the corridor as the bodies of the dead and injured lay piled high at the far end.

"You OK?" Pete asked Rhiannon.

"Fine, you?"

"I'm good. Alan?" said Pete.

"All good, wait a minute, how did you—"

"I'll explain later," said Pete before releasing the restraints and jumping uneasily to the floor. He paused as giddiness took hold, then breathed deeply until the sensation passed. He gestured to the injured. "C'mon, they need our help."

Soon Rhiannon and Alan were by his side, making their way to the injured. They joined the other crew members in pulling the survivors free and lining up the bodies of the less fortunate. Soon two triage teams arrived at the scene, pushing back many of the initial helpers.

"Attention crew," came a voice through the air. "This is a temporary stop to attend to the injured and assess damage. We will soon be returning to Earth to await further orders."

Pete backed away from the coal face and attempted to reach Cooper on the communicator – nothing. He tried The Ginge with similar results, then, finally, Knuckles.

After a few seconds he got an answer. "Pete! Where are you?"

"Knuckles! Oh thank god. We're on The Dorsano, back at Baltwae according to the HUD. Where are you guys?"

"Just me, I'm afraid. Gartorgia, Wendy's brothers pulled me from the wreck."

Rhiannon made her way to Pete. "What about Cooper and The Ginge?"

"I… I don't know. They survived the explosion, I know that. We became separated then comms went down."

Tears welled in Rhiannon's eyes.

Pete reached out to her, squeezing her hand. "We'll go back."

"Are you crazy?" said Knuckles. "That place will be crawling with Chardrekk by now."

Pete exchanged glances with Rhiannon. "Sounds about as friendly as the Dorsano."

"What?"

"I'll explain later, you want to join us?"

Knuckles looked out of the viewshield at the approaching Gartorgia, in the distance he could hear his brothers-in-law laughing. In an instant he knew there was no decision to be made. "Count me in."

"Brilliant, we just need to find a ship and we're out of here," said Pete.

"Surely there are other warp capable ships in the Baltwae loading docks," said Rhiannon.

Alan saw Pete and Rhiannon in conversation and left the rescue effort to join them. "Everything OK? I mean, apart from everything here... and the captain double-crossing us thing... and the missing boys... and... I'll shut up now," he said as Rhiannon glared at him.

"We're going to get out of here before Reynolds comes looking for us," said Pete.

"Good call. Where to... and in what?"

"Back to Traigh system in... we're not sure yet, hopefully there's something in the Baltwae docks… or maybe Knuckles scores something good," said Pete.

"From memory there's a gesthion class transit vessel and a 114 bariol diplomatic shuttle, still in the dock. Both are hyperspace capable," said Alan.

"How do you even know that," said Pete before pressing against the wall to make way for the first of the rushing rescue crews.

Alan and Rhiannon watched on from the other side of the corridor as a second team followed directly behind, carrying another hover crate loaded with injured. From the other side of the corridor two more teams arrived.

The three converged in the middle of the corridor again.

Alan was two shades of white lighter than seconds earlier. "Traigh? But we only just got out of there. And I mean just. And there'll be bogeys every—"

"Cooper and The Ginge are there," said Rhiannon.

"They're alive?"

"We... don't know."

"And you think it's a good idea to go back and help?"

"We think it's the only idea," said Pete.

Alan stared vacantly as he processed the situation. "But—"

"We're going, Alan," said Rhiannon.

"You can either come or wait here for Reynolds to come looking for us," said Pete.

Alan's expression hadn't changed from earlier, except perhaps his drooping bottom lip falling even further.

At the far end of the corridor a group of armed men appeared. Two men at the front scanned the busy corridor, spotted Pete and headed in his direction.

"They're onto us!" whispered Pete before turning in the other direction and heading back to the docking bay.

Although he was walking, Pete's pace was fast. Rhiannon and Alan struggled to keep up without breaking into a jog. Ahead the doors to the cargo bay were reopening.

Rhiannon turned her head back to the threat. "They're following us!"

Pete went to teleport then remembered that option was unvailable. "Get ready to run," he shouted.

Rhiannon looked at Pete's injury and Alan's everything. "You sure that's a good idea?"

"How else are we going to get to the docking bay?"

"I've got a better idea," said Alan.

"Freeze!" said one of Reynold's men as his group passed the injured and picked up speed in pursuit. They raised weapons at the trio.

The rest of the crew saw the situation unfolding and scrambled clear as a new kind of mayhem broke out in the corridor.

"Follow me!" said Alan as he scrambled through a side access door.

Pete and Rhiannon followed just as one of the soldiers yelled, "fire." Blasts whizzed by them as they left the corridor.

Once they made safety, Alan sealed the door behind them. Pete and Rhiannon looked around the new space – they were surrounded by pipes, service ducts and the inner workings of a starship they'd never seen.

"Where are we?" said Rhiannon.

"Corridor MS38b," said Alan as he jumped onto a ladder and headed down. "Follow me!"

*

The stench of burning chemicals filled the air as members of the ground crew were busy battling the fire that engulfed several craft near the entrance. The entrance itself was well on the way to being fully reopened.

A small thumping sound thudded out from the floor, a mere side note in the craziness that was the emergency in the docking bay.

"Stupid thing," said Alan, as he thumped the access hatch.

"What's going on?" said Pete.

Alan hit the hatch again, lifting it. The light of the docking bay flooded in, then he pushed the hatch clear. "Oh yeah!"

He popped his head through the gap and scanned the space. "Boom! Knocked it out of the park!" he said as he hoisted himself onto the docking bay floor. "Quick, this way."

Before long the three were standing in front of their ship once more. The stairwell leading to the cockpit was still open and Pete ushered Rhiannon, then Alan, up. He climbed to the top of the stairs himself before casting an eye back over the docking bay just as Reynolds' men entered via the corridor.

They spotted him, but Alan's shortcut had given them all the head start they needed.

By the time he reached the cockpit the others were strapped in and bringing the ship to life. The engine hummed.

Alan consulted visual displays. "We're showing some damage—"

"Can we fly?"

"Erm… I think so."

"That'll have to do," said Pete as he lifted the gunship as delicately as he could between the craft it was wedged between. They raised clear of the parked fleet. Through the viewshield the approaching men could be seen giving up their foot chase and consulting superiors on their communicators to plan the next move. Pete gave them a broad smile then a hand gesture that could have resembled rolling dice or another, more personal, male pastime.

"Docking bay to gunship zero two, identify," came a voice through the ship's comms.

Pete rotated the vehicle around to face the exit port as he and the others exchanged glances.

"Docking bay to gunship zero two, all ships are grounded, identify immediately."

Pete picked up his communicator. "Docking bay, this is Pete Bates."

"Sir, you cannot take that ship. We are in a total no-fly situation, with departure to Earth imminent."

"You'll be a ship short, I'm afraid; we're heading back to find the captain."

"Sir?"

Pete locked the ship in line with the exit and pushed it forward with his controller.

"Sir, with respect, Captain Reynolds has specifically said no one is to leave the ship."

"Firstly, it's the acting captain; secondly, he can blow it out his arse… with all due respect."

Pete hammered the ship at pace over the past small distance to the exit, then banked hard right when he burst into open space. "Whooooo-hooooooo!!"

Rhiannon and Alan laughed.

"I can't believe you just told Reynolds to blow it out his arse!" said Rhiannon through tears.

"It was the best I could come up with a short notice."

***

CHAPTER 36

The Ginge looked at Edson Pelk. "So, and I just want to get his 100 percent correct in my head, when you say life forms, you're not talking about microbes and stuff, you're talking about animals, like predators, bigger than this thing," he said, pointing to Gluff Hurn, who responded with a look that made him aware not all the predators were outside of the gunship.

"No offence," said The Ginge after a large gulp. Then he refocussed on the visual displays of the formidable local wildlife. "And they're basically hiding in the caves, waiting for their chance to eat us?"

There was a thumping on the hull.

"Sorry, did I say hiding? Scratch that."

"We are perfectly safe within the ship," said Dilania.

"For how long?"

"For as long as—"

The thumping noise echoed through the cockpit again, this time followed by a warning alarm on the ship's displays.

"Captain, we're taking some damage to the drive systems."

"Acknowledged, Edson."

"Awesome," said The Ginge. "Can't Gluff jump on the turret and blast them?"

"It would be unwise to fire in such uncertain geological circumstances."

The Ginge groaned.

"So, how long do we need to stay here?" said Cooper.

"We'll need to wait for the Chardrekk craft to move in, clear the area of debris, then dissipate," said Dilania. "And, of course, repair the ship to full flight capability."

There was another thud at on the hull, this one from a different location. It too was followed by another warning alarm.

"Awesome," said The Ginge. "Just awesome."

*

Alan looked at a new alert on his display holos. "Pete, incoming holocomm, it's Reynolds."

Instantly smiles were wiped from the faces of Pete, Rhiannon and Alan, their gunship suddenly feeling not such a safe place to hide under the eyes of an upset captain aboard a starship.

Pete swallowed hard and shared unsure glances with the others.

"I'm going to take it."

"What if he wants to kill us," said Alan.

"At least we'll know about it in advance."

"And that's a good thing?"

Pete ignored Alan, stood, then cleared his throat in preparation. "Punch him up."

Reynolds hologram appeared in the cockpit in front of the three. The Dorsano's new captain studied them before fixing his eyes on the pilot.

"What?" said Pete, with an unsure amount of defiance.

"Where do you think you're going?"

"To see if the actual captain of your ship is still alive."

"We have orders to return to Earth."

"Who's orders?"

"I am not interested in discussing complex battlefield strategy with some spotty teenager, who has no right to wear the uniform."

Pete locked eyes with the hologram. "Someone who goes around killing their own kind to get ahead has no right to wear the uniform."

To his right, Pete could see Alan making a hand signal, pleading with him to not rile the acting captain further.

Reynolds' image seethed. "It is not clever to comment on issues you know nothing about."

"In that case, why don't we leave the fleeing to you and you leave the rescuing to us. That way everyone's happy."

There was a tense pause as Reynolds thought on the proposal and Pete prayed he and his crew would be let on their way.

"A fair deal," said Reynolds.

Pete nodded to the hologram and shared subtle smiles with his crew. "Thank you, acting captain."

"Oh, one more thing," said Reynolds. "You may also be interested to know that Baltwae has served its purpose. It holds valuable secrets and, as such, is scheduled to be destroyed."

"What? When, exactly?"

"Now," said Reynolds before turning to his crew. "Rodriguez, destroy the Baltwae base."

"What?" screamed Pete. "You can't!"

Reynolds looked back at Pete; a warm pleasure sparkled through his eyes. "I guess I'll leave you to the rescuing then. Good luck."

"Reynolds! Rey—" screamed Pete.

But it was too late. Reynolds image vanished from the cockpit. Seconds later, the Dorsano unleashed a wave of furious fire, targeting on the small Baltwae outpost that had been their home for months.

Explosions ballooned out into space. Pete fell to his knees, Rhiannon wailed and Alan stared on in stunned bewilderment.

"Nooooo!" wailed Pete.

*

"Captain, the first enemy ships are entering the atmosphere," said Edson.

Dilania looked at the holodisplay, showing a squadron of attackers spreading out across the sky and diving closer to surface level. "Acknowledged, running systems light."

"Confirmed."

The watched the ships dance across the screen in silence, as if breathing too loud would give away their position. Cooper and The Ginge exchanged expressions of concern across the cockpit, not just at the current situation.

The Ginge mouthed something to Cooper, who scrunched up his face and mouthed 'what' in return. The redhead repeated his initial silent statement but, once more, it was too complex for Cooper to comprehend.

Then Cooper experienced a sensation he hadn't felt since his time on Galactica –The Ginge was entering his thoughts. He prepared a lobby in his conscious mind for the conversation to take place in.

'Cooper, it's me,' thought The Ginge.

'Yeah, I figured that. What do you want?'

'What's the plan?'

'I don't have one, you?' thought Cooper.

'I've got nothing, but we need to get in touch with the others.'

'I'd rather they stayed with the Dorsano and tried to figure out what's going on.'

'Really? If Reynolds actually did get rid of us, who do you think would be next?'

Cooper's mind raced with thoughts and possibilities, all of which he kept from his thought conversation lobby.

'Cooper?'

'Yeah?'

'Oh, nothing, I thought I'd lost you.'

'Nuh, just thinking.'

'We have to get them on comms.'

'I'm not sure if that's safe for them or us right now.'

'So, what then?'

'Bide our time here. When we get out we can work out our next move.'

'Unless the Chardrekk get us first, or we're eaten by whatever freaky alien killing machine is probably chewing their way through the hull as we speak.'

'We'll be fine.'

'Coops, I'm in your thoughts, I know you didn't even believe that. Besides, even if we do get out of here, where the hell do we go?'

'The Dorsano.'

'News update, dude – that's not your ship anymore.'

'It's still my ship.'

'Well, I'm not entirely confident Reynolds will be happy to see you.'

'There's always plan B.'

'Plan B?'

'The other starships we're building.'

'Yeah, I think Oil Man might have a problem with that.'

'Not those starships, the other ones.'

'What other ones?'

'The insurance policy ones.'

'What are you talking about?'

'Just a little side project I'm working on with Walter.'

'Our own science teacher Walter? What, you just started building your own starships?'

'Yes.'

'Genius, Coops, pure genius.'

*

"Go easy," said Alan as he consulted the displays of the ruin of the Baltwae docking bay. "Nice and easy. Left! Left!" he screamed.

Pete jumped from the shock of Alan's vocal escalation and the gunship wobbled before Pete regained control. "It's a bit difficult to go easy when I'm sitting next to panic central."

"Sorry… this is tense," said Alan as he wiped the sweat from the palms of his hand on his uniform.

"Focus boys," said Rhiannon. "There's a good spot to land just the other side of this wreckage. Just ease through the smoke, it's safe."

"Only if Alan doesn't get his knickers in a knot."

Alan went red, but nobody noticed as Pete steered the gunship through the billowing black smoke and Rhiannon consulted telemetry and displays. The viewshield went dark for several seconds, then released its choke hold, showing the remainder of the damaged docking bay and the first few glimpse of the pummelled infrastructure.

"How could he do this?" said Rhiannon.

"Makes me sick," said Pete as he brought the ship to the ground.

"You guys see anything flightworthy?" said Alan.

"No, you?"

"No."

"So, what, we're stuck here?"

"Until we find a lift, looks like it."

*

Far beyond the gas giant Zoesam in the Traigh system, a piece of alien spaceship spiralled toward the darkness, nothing slowing it from the explosions that hurled it into the black.

This was no ordinary piece of wreckage. The large mass, encased in blast-proof material, had survived a collision between two starships, keeping its structural integrity intact and its occupants alive.

While the standard crew of the Cremmerson suffered a far more instant and permanent fate, the senior crew on the protected bridge simply rode the blast wave. It was several hours before the first of the rescue craft successfully tracked their distress beacon to reach their location.

The rescue ship manoeuvred into position, matching the rotation of the cockpit module, before an arm reached out, connecting the vehicles. A walkway was deployed, enveloping the hatch of the craft at either end. Soon, the hatches were opened and the crew from the bridge of the Cremmerson made their way to safety aboard the rescue craft.

Isla Enchant felt the humiliation hit her at a new level as each phase of the rescue passed; it hit a crescendo when she sat alongside her subordinates being ferried back to Traigh. Through all the hubbub of activity around her she did not speak. When her rescuers tended to her, when she passed the scene of the battle, clean-up already well under way; she did not speak.

When she entered Traigh's orbit, however, her silence was broken. She turned to her immediate underling. "I want to know the identities of every raider who survived."

"Yes, sir."

"Then we are going to destroy them."

*

Reynolds saw his communicator buzz into life, saw the caller and swallowed heavily before responding. "Sir?"

"What is your progress?"

"We are back at Baltwae, sir. Preparing the jump to Earth."

"And the former captain?"

"MIA, sir."

"Missing? That's not as permanent as I'd expected. And the others?"

"Either MIA or heading back into the warzone to find them, sir."

Oil Man breathed heavily through disappointment. "Very well, return to Earth. Meanwhile, I shall tidy up the loose ends."

Reynolds felt the mark of failure cast upon him. In the face of missing expectation he lifted his chin and puffed his chest out, ready to fight his next battle. "Sir, yes, sir."

*

"Grand Leader, High Commander Isla Enchant is landing shortly. Are you prepared to greet her?"" said one of the servants to the Chardrekk inner council.

Entax Pivak remained seated, not breaking his view of the Traigh skyline – a vista of shuttle craft and gleaming towers climbing into the rich purple skies. "I believe the High Commander can make her way to me, when she has mustered the confidence."

"Understood, Grand Leader."

The servant retreated from the scene, leaving Pivak to his thoughts and the view until his communication alert bleeped into life.

"Yes?"

"Entax Pivak, thank you once again for your time, I have further information you may be interested in," said Oil Man.

"Be quick, human."

"The fate of the humans Cooper Simpson and Wesley Ellis is not known. Alive or dead, it is believed they are still in the Traigh system. The others, Bradley Thomas and Pete Bates plan to return shortly to search for them. They will soon all be within you reach, Grand Leader."

"Excellent," said Entax Pivak.

"…and our arrangement?"

"What of it?"

"I'm referring to my Lore sanctioned leadership of the Earth, when war is done… I mean the deal we stru—"

"Do not embarrass yourself on the topic any further. Our deal is set… once those humans are mine."

"Excellent," said Oil Man. "I can assure you I will do everything in my—"

Entax Pivak closed the communications connection to be alone with his silence and skyline.

*

Pete, sat with Rhiannon in what used to their choice bar in the Baltwae outpost. The air was thick with the putrid scent of destruction as the waft of fire and fumes soured every breath. Alan shuffled back to the table through the broken furniture and debris then placed three glasses and a bottle on the table in front of the others.

"This was the only thing I could find," he said as he upturned a nearby chair and tested its stability. "So, we're drinking... Bwella drops apparently."

The others didn't speak, defeated by circumstance and loss. Alan dutifully filled the glasses and placed them in front of Pete and Rhiannon. "C'mon, drink up!"

After failing to motivate the others into indulging, Alan decided to lead from the front, taking a large swig from his glass. He quickly realised his error and gasped for fresh air as the strong alcohol burned his underprepared throat. "Phew, got some kick to it," he said as his eyes moistened.

He collected himself and studied the others, both unmoved from his antics. "Dammit, will you two say something?"

"Like what," snapped Rhiannon.

Alan reeled back from the response but bravely, well, for him, stared back at her. "I don't know, but sitting here feeling sorry for ourselves isn't going to help us and it sure isn't going to help Cooper and The Ginge."

"If they're alive," she said before bursting into tears.

Alan took another, less substantial sip of his drink for courage, then leaned over to Rhiannon and offered an embrace. She looked him up and down in suspicion before giving up and burying herself in his arms, wailing into his uniform.

"It's alright," said Alan as held her. "It's alright."

Pete didn't know whether to watch on or look away; instead he reached out for his drink and took a gentle shot. The drink hit him like a wall. His swallow mechanism and gag reflex battled to decide the future direction of the beverage. But Pete wasn't getting beaten now; he swallowed, took a few deep breaths then backed up with another shot.

He reached inside his uniform and pulled out the small metallic cylinder he kept stashed there. He held The Ginge's prized bullet, spinning and toying with it in his hand. Then he thought about his redheaded friend and the look of horror on his face when the Jack Reacher round situation happened. He laughed. Then he thought of Cooper, then the two of them missing.

Maybe it was the emotion, or the helplessness or the bwella drops but the image of his two friends was so powerful in his mind – so strong. It was as if he could reach out and grab them.

He took another swig, his awareness of Rhiannon and Alan now a distant blur. He gripped the bullet and focused his thoughts to Cooper and The Ginge. So real now, their presence almost with him. Somehow, he knew they were alive – he just knew.

Then, just when he was starting to question his sanity he felt The Ginge calling his name. No, he wasn't calling it, he was thinking it. The Ginge. Thought sharing. It was distant, so very, very faint and distant but it was real. 'Pete, Pete, you there? Pete?'

Pete lowered his head to the table, closed his eyes and blocked his ears so he could clear all non-essential sense inputs. 'Ginge?' he thought.

'Pete! Coops, did you get that? It's him!'

'I got him! Hey, Knuckles, you still there?'

'Unfortunately.'

'Where are you guys? How did you…'

'Survive? Dilania and her crew got us, when they were shutout from the Dorsano,' thought The Ginge.

'We're still on Hazsis, holed-up in a cave,' thought Cooper. 'The place is teeming with Chardrekk. We're laying low until the heat dies down. Then we're going to repair the ship and see if we can't steal ourselves a little upgrade.'

'Grand Theft Starshipo,' added The Ginge. 'We'll haul ass before they have five wanted stars on us though. Gangsta.'

'Alright Ginge, focus.'

'Sorry.'

'We can come and get you,' thought Pete.

'Negatory, we can't risk everyone in the same place. Besides, we need you aboard the Dorsano to—'

'We're no longer on the Dorsano.'

'What?'

'Reynolds went psycho, I'm pretty sure he wanted us dead so—'

'Rhiannon… is she?' thought Cooper.

'She's fine. Her, Alan and I swiped a rescue ship and headed back to Baltwae.'

'Baltwae… we could make that work, we just—'

'Reynolds blasted the crap out of it, there's barely anything left.'

'What an ass-hat!' thought The Ginge.

'Says he's following orders,' thought Pete.

'Orders?' thought the Ginge. 'It's gotta be Oil Man – ass-hat two point zero.'

'But why?' thought Cooper, 'Betraying us at the expense of the New Council makes no sense. Not in isolation, not even for them. No, there has to be something else, some connection we don't know about.'

There was a pause in conscious thought as the four contemplated what it could all mean.

'We'll figure it out,' thought Cooper, eventually.

'So, what do you want us to do in the meantime?' thought Pete.

'What state is Baltwae in? Salvageable enough to get the docking bay up and running again?'

'Maybe. Why?'

'We're going to need a few people we trust if we're going to fill a starship,' thought Cooper.

'The only two problems I can think of with that is we don't have any people we could trust or, you know, a starship.'

'After this I want you to get in touch in Walter. Update him on what's going on and tell him I've told you about his special project.'

'What special project?'

'The one where he's secretly building a starship just in case everything went pear-shaped with the Dorsano,' thought Cooper.

'Seriously? Nice! And the crew? I don't know who you think you can trust right now but my list extends to the four of us Rhi and Alan.'

'Which is why I need you to get in touch with Terenda.'

'Ahhh!' thought Pete.

'…and ask her to get in touch with the survivors of the new intake army from back on Galactica.'

'Ahhh!'

'They're the only people we can get close to trusting right now.'

'We're getting the band back together,' thought The Ginge.

'Genius,' though Pete, excited at the idea. 'What about the whole Terenda/Rhiannon thing? Not going to be a problem?'

'She'd have to get past the Dilania/Rhiannon thing first,' thought The Ginge, before reflecting on his failing reputation. 'So typical, life gives me five seconds in the spotlight, before normal broadcasting resumes and I'm sitting in the stands watching Cooper deal with his oversupply problems.'

'Focus, Ginge!'

'That's easy for you to say, you haven't gone from Hugh Hefner to a monk in six months.'

'Did you just compare yourself to Hugh Hefner?' thought Knuckles. 'Wow!'

'What's so wrong with that, I could easily ha—'

'Can we just focus for a minute?' thought Cooper. 'I don't know about you but this thought-sharing is starting to wear me out. I'm not sure how long I can keep this up.'

There was a pause as the boys re-railed their train of thought.

'Good,' continued Cooper. 'So, Pete, get onto Walter, get an ETA on the second ship and prepare Baltwae for some old friends. Doable?'

'Doable.'

'Knuckles, get your hands on a hyperspace worthy ship and join Alan and the others as soon as you can.'

'Affirmative. The sooner I'm out of Gartorgia the better anyways.'

'The ex?' thought The Ginge.

'No.'

'Kids?'

The others sensed Knuckles tension, what they didn't detect was his thoughts turning to Wendy's brothers and the humiliations he felt being indebted to them for rescuing him once again. Alone it was enough to churn the pit of his stomach and when coupled with the unsanitary state in which he was rescued and the fact they would never let him forget it, it sickened him. 'It's complicated, alright.'

'OK then,' thought Cooper.' Meanwhile Ginge and I will get our asses back to safety, then we'll plan sweet revenge. Deal?'

The others agreed.

'You know what this all means, don't you?' thought Pete. 'It looks like we were making part two of a trilogy all along.'

'True,' thought Cooper, his feelings of bittersweet irony successfully being translated to the others. 'I hope things end better in part three.'

'That's the plan.'

That was it. With their directions laid out, they kept the link open as long as they could, sharing thoughts before the exhaustion made it all too hard to continue. They communicated around the edges of their thoughts; the small stuff. As if any further discussion on what was about to unfold would bring up emotions too big to calibrate.

They were all isolated, in their own way. They were all doubtful on the credibility of the plan and the chances of them ever seeing each other again, let alone fulfilling their mission. Their hearts had all grown harder with the pain and loss of battle and betrayal. But, for those moments, before the drain of thought-sharing became too much to take, teasing The Ginge about his poor form with the ladies, Pete about his swearing policy, Knuckles about his family connections or Cooper about his impending complicated love life seemed to make none of it matter. For just that instant in a lifetime, they weren't four warriors set on a course they would never choose, dictated by forces greater than they could understand and destined to alter them forever. They were simply four teenagers, being.

***

THE END

###

COMING 2027

****

ALMS FOR
AN AUTHOR

Your reviews matter big-time to this indie author. If you can post your thoughts on Amazon.com or Goodreads.com it would be greatly appreciated.

Review on
Amazon.com

Review on
Goodreads.com

****

ABOUT THE AUTHOR:

Like the legendary R M Williams, Matt was born in Jamestown in rural South Australia. But that's where the remarkable similarities between these two end. While Reginald went from bushman to world renowned millionaire outback clothing designer, Matt is a complete dag who was lured by the city lights of Adelaide. Kindergarten in the big smoke was a culture shock, but it is here he first discovered his love of storytelling.

In high school that love found an outlet in a series of completely unflattering cartoons about fellow students and teachers alike. He survived long enough to further his art into a successful career in multimedia design but, like a zombified leech, the lure of the written word gnawed at him, forcing him to pen his first novel, the award-winning sci-fi comedy epic, Kings of the World. It was followed the next year by Amazon Australia dystopian sci-fi best-seller Apocalypse: Diary of a Survivor.

Matt donates part-proceeds of each book sold to find a cure for Rett Syndrome, a neurological condition the youngest of his three children, Abby, has. As a gorgeous Rett angel, Abby cannot walk, talk or use her hands in a meaningful way. So, not only is each of your book purchases a ticket to fantastically rounded, character driven, hilarious and poignant sci-fi awesomeness, it wraps you in a warm feeling that you've made a difference to people who deserve your help the most. Like the zombified leech it's a no-brainer.